BRAIN FOOD

HOW TO EAT SMART AND SHARPEN YOUR MIND

Dr Lisa Mosconi

PENGUIN LIFE
AN IMPRINT OF
PENGUIN BOOKS

PENGUIN LIFE

UK | USA | Canada | Ireland | Australia
India | New Zealand | South Africa

Penguin Life is part of the Penguin Random House group of companies
whose addresses can be found at global.penguinrandomhouse.com.

First published in the United States of America by Avery,
an imprint of Penguin Random House 2018
First published in Great Britain by Penguin Life 2018
This edition published 2019
001

Neither the author nor the publisher is engaged in providing any
medical diagnosis or treatment to individual readers. The information contained
in this book is no substitute for medical advice. If you have any specific questions about
any medical matter you should always consult your doctor. Neither the author
nor the publisher can accept any liability for any loss, injury or damage
which is sustained by readers in consequence of adopting any suggestions
or of using the information in this book

Printed and bound in Great Britain by Clays Ltd, Elcograf S.p.A.

A CIP catalogue record for this book is available from the British Library

ISBN: 978-0-241-38177-9

www.greenpenguin.co.uk

To my family,

whose love I shall never forget

CONTENTS

STEP 1
UNDERSTANDING NEURO-NUTRITION

STEP 2

EATING FOR COGNITIVE POWER

STEP 3

TOWARD THE OPTIMAL BRAIN DIET

PREFACE

A few years ago, I delivered the keynote at an international conference on the prevention of Alzheimer's disease. It was a beautiful sunny day in Italy, the lecture hall brimming with doctors, students, and laymen, all eager to hear the latest on pharmacological treatments for Alzheimer's.

I was much less eager to be the bearer of bad news. Unfortunately, current medications for Alzheimer's lessen symptoms for a limited amount of time but cannot stop the damage that aging and disease cause to brain cells. A new generation of disease-modifying drugs is under development, but clinical trials have yielded mostly disappointing results so far, confirming what everyone knew: there is no cure in sight.

At which point someone in the audience asked: "How about olive oil?"

My neuroscience-trained brain was baffled. *Olive oil?*

Olive oil was not in any of my research proposals, or in any way part of my education. I got my PhD in neuroscience and nuclear medicine to focus on the genetic aspects of the disease, motivated in part by seeing its devastating effects on my immediate family. My work for the last fifteen years has been focused on the early detection of Alzheimer's. Specifically, in my research I use brain imaging techniques, such as magnetic resonance imaging (MRI) and positron emission tomography (PET), to look at people's brains in relationship to their genetic backgrounds, and in doing so, learn about their likelihood of developing disease.

This work led me to direct the Family History of Alzheimer's research program at the NYU School of Medicine back in 2009. The program focuses on the children and family members of Alzheimer's patients. Everyone there has broadly the same concern: "Am I at risk for Alzheimer's and what can I do to make sure I don't get it?"

Over the years, I experienced a change in the kinds of questions I was asked by our patients, much like the olive oil question at the conference. Beyond the discussions about genes and DNA, sooner or later the conversation turned to food: "What should I eat to keep my brain healthy?"

While all my research draws on my education as an adult, everything I associate with food comes from my upbringing in Florence, Italy. Back in my hometown, I developed a heartfelt appreciation for wholesome, healthy food from a very young age, which I took for granted until I moved to the United States to study for my PhD. I had not anticipated the challenge of finding a flavorful tomato or the artery-clogging danger hidden within a seemingly innocent chocolate chip cookie. As I struggled with my own diet in my new environment, I learned from my research that I was not alone. By their own accounts, more than half of my study participants reported only trace amounts of vegetables and fruits in their diets.

Little by little, it became clear that I wasn't just far from home. I was also far from the original theses about the genetic aspects of dementia. In fact, it turns out that the role of genetics in Alzheimer's, and dementia in general, is not as major as we previously thought. While some patients carry aggressive genetic mutations that cause dementia, for the vast majority of the population, risk is influenced by a variety of medical and lifestyle factors—including a person's diet. When my research revealed how important, and how oddly neglected, diet and nutrition had been in the field, I went back to school and completed a third degree in integrative nutrition. I drew upon that and other work to found the Nutrition & Brain Fitness Lab at NYU, with the goal of identifying lifestyle factors that support the brain's health, safeguarding it against dementia. A few years later, I initiated the Brain Nutrition coursework and began teaching at the NYU Steinhardt School's Department of Nutrition and Food Studies. Around the same time, I moved to Weill Cornell Medical College, where I have the honor of serving as the Associate Director of the first Alzheimer's Prevention Clinic in the country. The clinic's innovative approach includes pharmaceutical as well as behavioral interventions aimed at improving both medical status and lifestyle choices in an Azheimer's-preventative way. Diet and nutrition are a big part of the practice. Altogether, this work has led me to dive headfirst into the complex relationship between our brains and the foods we eat, and to educate the public as to how to eat healthfully for their brains.

As with anyone who has ever gone down a dietary path toward optimal nutrition, I quickly realized that the available advice was often conflicting and incoherent. But as a scientist, I was above all surprised at the volume and effect of pseudo-scientific information available on the Internet, especially in comparison with how little had been published with the rigor of peer-reviewed medical journals.

We've heard a lot about what is and isn't good for our brains. For example, many of us have recently become aware of an American gluten panic. But only a few years ago, grains were considered the epitome of eating healthy—and people were terrified of eating fatty foods. The problem is that while many of the recommendations you can find online promote a scientific worldview, very few have been substantiated by sound research. The Internet and media in particular have a tendency to make broad extrapolations on limited findings and sensationalize everything. For instance, at least once a week someone will ask my opinion on the latest "miracle drug" for Alzheimer's. I'll look up the study. More often than not, it's true that the drug worked . . . but in a population of ten mice. So that's great news if you are one of those ten mice. Whether or not those findings are relevant to *people* is a whole different story.

This is where scientific literacy comes into play. Which sources of information are the most credible? How can we know if the study we heard about in the evening news is worth acting upon?

While sound scientific research is more limited than Internet blogs, results do prove to be consistent. A new generation of studies has begun to identify which nutrients are particularly helpful in enabling our brains to work at the max, as well as protecting them as we age, thereby granting us continued mental strength over the course of a lifetime. At the same time, we are also learning which nutrients are harmful to the brain, negatively affecting our cognitive abilities and increasing the risk of mental deterioration. This includes my personal experience accumulated over years of hands-on research regarding the important interactions among genetics, nutrition, and lifestyle.

What's important to note is that I'm not just presenting the results of my own research, but rather the analysis of hundreds of scientists who have been studying the relationship between the foods we eat and the health of our brains over the decades. I'm hoping to make clear that science is never a result of one person's opinion, but always involves a collective of doctors, scientists, and even you, the population itself, in an ongoing, educational exchange that unfolds over time. Thanks to this worldwide dedication, we challenge one another to get to the bottom of those things we're most determined to resolve. In fact, the beauty of science should be strength in numbers.

The danger of looking at isolated papers is that you might find that what appears to be fact in one study is proven wrong in the next. One day you read that "according to science" you must avoid cholesterol at all costs. Then you turn around to find another equally "scientific" explaining the role cholesterol plays in supporting a healthy brain. How can both things be true?

Ultimately, no single study is perfect. One can never be absolutely sure that its findings are valid and applicable to the population at large. We need to look at a bigger picture. The more an independent study replicates a specific finding, and insists upon applying a wide range of methods to an equally wide number of patients, the more likely that the finding is actually true and applicable to everyone.

Make no mistake. There is a bottom line when it comes to what's good for your brain and what's not. In *Brain Food*, I draw on my background as a neuroscientist to build a neurological and nutritional framework around the ways that food is specifically vital in promoting optimal brain health. In the pages that follow, we will get down to the details of what science has discovered so far by exploring *neuro-nutrition*, or *nutrition for the brain*. We'll look at how food breaks down into nutrients, and to what degree these nutrients feed our brains. We'll talk about how the brain actually works, and the specific influence diet has on our cognitive performance. But mostly, we'll see how the human brain has its *own unique diet*, different from that of the rest of the body. Just as we would eat differently to slim down than to train for a triathlon, when optimizing for long-term cognitive health, the brain has its own demands. As it turns out, our future lies in our own hands—and what's on our menu.

STEP 1

UNDERSTANDING NEURO-NUTRITION

1

The Looming Brain Health Crisis

THE GOOD NEWS

Let's start with some good news. We, as a human race, are living longer than ever before. Life expectancy has been steadily on the rise for over two hundred years. During the twentieth century in particular, there has been nothing less than a downright boom in human longevity. This dramatic increase in life expectancy ranks as one of society's greatest achievements. According to the Centers for Disease Control and Prevention, while most babies born in 1900 did not live past age fifty, life expectancy now averages just under eighty years old in most industrialized countries.

It turns out that the secret behind our recently extended life span is not due to genetics or natural selection, but rather to the relentless improvements made to our overall standard of living. From a medical and public health perspective, these developments were nothing less than game changing. For example, major diseases such as smallpox, polio, and measles have been eradicated by mass vaccination. At the same time, better living standards achieved through improvements in education, housing, nutrition, and sanitation systems have substantially reduced malnutrition and infections, preventing many unnecessary deaths among children. Furthermore, technologies designed to improve health have become available to the masses, whether via refrigeration to prevent spoilage or systemized garbage collection, which in and of itself eliminated many common sources of disease. These impressive shifts have not only dramatically affected the ways in which civilizations eat, but also determined how civilizations will live and die.

In the end, we are living longer and longer lives. In most industrialized nations old age is now a reasonable expectation, so much so that scientists are adamant: an older society is here to stay. That's good news—news that is hard won over the millennia of the history of humankind.

THE NOT SUCH GOOD NEWS

Now for the flip side. As it turns out, to some degree, we might be victims of our own success. Unfortunately, this increase in life span has not necessarily provided us with additional years of particularly high-quality health. Old age can come with wisdom, but it just as regularly arrives with some less illustrious additions. Hearing loss, bifocal glasses, slower reflexes, and common medical ailments such as arthritis, rheumatisms, and respiratory problems are examples of those side effects we'd rather do without. What is of greater concern is that deterioration of the brain sneaks up on many of us as we age, making us vulnerable to memory deficits and loss of cognitive function.

Over the years, I've asked countless patients, "What concerns you most about your future health?" More often than not, it wasn't the condition of their heart or even the risk of cancer that came to mind. Today, the greatest fear for most people is that they might end their days battling dementia.

The most common cause of dementia, and probably the most feared, is the memory-robbing Alzheimer's disease. The idea of losing track of one's own thoughts, or being unable to remember our loved ones, is cause for great anxiety, fear, and stress. Equally daunting is our inevitable grief at seeing a relative or close friend suffer from this devastating disease.

This concern is understandable. Of all the challenges to aging in the twenty-first century, nothing compares to the unprecedented scale of Alzheimer's. According to recent reports from the Alzheimer's Association, the number of people living with Alzheimer's in the United States alone is an estimated 5.3 million. As the baby boomer generation ages, the number of patients is predicted to reach a staggering 15 million cases by 2050. This is the population of Los Angeles, New York, and Chicago all together.

A similar trend is observed planet wide. Today more than 46 million people live with dementia the world over. This number is estimated to increase to 132 million by the year 2050.

Further, while Alzheimer's represents the most recognizable (and most common) framework for dementia, there are many ways a healthy brain can go awry: other forms of dementia, Parkinson's disease, stroke, depression, and so forth. As more countries reap the benefits of longer lives, the burden of all these disorders is reaching an alarming proportion. If that weren't enough, beyond specific disorders, general age-related cognitive impairment might affect three to four times as many people, with extraordinary psychological, social, and economic consequences.

As we take in the challenges of such an unprecedented brain health crisis, the year 2050 doesn't seem so far away.

We need a cure, and we need it fast.

THE BREAKING NEWS

Now for the news that provides us with hope. Recent medical breakthroughs have radically changed our understanding of aging and disease by showing that the brain changes leading to dementia unfold over *decades* before anyone ever forgets a name or loses their keys. These findings have revealed a much more complex picture than previously imagined.

Two technologies in particular have deeply changed the way we understand brain aging. On the one hand, we finally have access to "cheap genomics" (affordable DNA testing), which allows us to take an important peek into our genetic predispositions. While just five years ago we would have had to spend thousands of dollars to do a proper genetic screening on patients, today anyone can obtain such precious information for just a few hundred.

In addition, we have lab tests such as brain imaging that allow us to view how the brain is functioning over time, in response to both our genetics *and* our lifestyle choices. Scientists now have access to sophisticated brain imaging techniques, such as magnetic resonance imaging (MRI) and positron emission tomography (PET), which allow a view of the human brain from the inside out. Brain imaging has given us a rare and opportune window from which we can catch a glimpse of the actual progression of many brain diseases years in advance of any noticeable clinical symptoms. Finally, we can track the development of diseases like Alzheimer's as they unfold, and use that knowledge to identify people at risk many years, if not decades, before clinical symptoms emerge.

As you'll notice, a large part of the discussion around nutrition for brain health references Alzheimer's. This is primarily because Alzheimer's is one of the few neurological diseases reaching epidemic proportions that scientists can agree is influenced by diet, and as such, it is the instigator for much of the funded research in the space. In order to figure out what people need to eat to improve or maintain optimal cognitive capacities, we need to compare people who age gracefully (from the brain's perspective) to those who unfortunately do not. In this context, Alzheimer's is in effect shorthand for the most extreme responses of the brain to the nutrients we provide. The lessons learned, and behaviors to follow, apply therefore to broader cognitive health as well as many, if not all, forms of cognitive decline associated with brain aging. In much the same way that following the guidelines to prevent heart disease is good for everyone—not just those at risk for cardiac events—the newly discovered dietary strategies to prevent Alzheimer's are also those that optimize *cognitive health overall*, over the course of a lifetime and with benefits across the board. Research findings in Alzheimer's can then be used as a framework that stands in for cognitive decline of the aging brain as a whole.

By using brain imaging, several teams across the world have been successful in mapping the development of Alzheimer's over time, showing how it occurs gradually in the brain and progresses over a twenty-to-forty-year period *before* clinical symptoms emerge. In other words, cognitive impairment is not a mere consequence of old age, but rather represents the endgame of years after years of accumulated insults to the brain. What's even more disconcerting is that the brain changes leading to dementia can begin as early as young adulthood, and in some cases, even from birth. As it turns out, Alzheimer's is *not* a disease of the old, nor does it hit without warning.

Currently, our understanding is that many genetic, lifestyle, and environmental factors can potentially damage the brain while one is still young, triggering a cascade of pathological events that ultimately lead to cognitive deterioration. Whether we're referring to the somewhat typical forgetfulness and mild memory issues that many people experience around age sixty, or to the full-blown dementia and loss of independent function in older age, there is a long period of time during which brain changes can be under way without the disease yet causing any noticeable symptoms.

If this sounds frightening, take heart.

The key message from these studies, including my own work, is that this lengthy gap leaves a precious window of time to finally and thoroughly explore the power of *prevention.* There is increasing evidence that implementing the lifestyle changes described in this book has the potential to prevent Alzheimer's from developing and also to help slow down or even halt progression of the disease in those who are currently suffering from dementia.

If that weren't enough, eating for your brain isn't just a powerful preventative against disease—it actually helps you achieve peak performance *in every part of your life.* Beyond the specific fears over any particular disease and toward a more general hope for better brain health over a longer life, this is a call to action. Anyone who is old enough to consider how their brain will remain healthy into old age is old enough to start making vital changes to address that immediately.

SIMPLY IRREPLACEABLE

Taking care of our brains is a lifelong process, mostly due to the very nature of our brain cells. In fact, our brain cells, or neurons, are literally *irreplaceable.* This is a major difference between the brain and the other organs in our body. In the rest of the body, cells are constantly replaced (think about how fast your hair and nails can grow). But the brain lacks the capacity to continuously grow new neurons.

While some neurons do continue to grow as we age, the vast majority stay with us for a lifetime, rendering them particularly susceptible to the wear and tear that naturally occur as part of the aging process. This is why diseases like Alzheimer's are so devastating—they deliver the final uppercut to neurons that can't be regrown.

All of this means that we need to pay even greater attention to the health of our brain cells, since, by and large, they might very well be the only ones we will ever have.

This is especially important given the limited insight we have as to what's going on inside our brains. Often enough, we don't even realize that our brains are suffering until an interior problem grows severe enough to produce external symptoms such as loss of consciousness, hallucinations, or cognitive deficits. For example, it is not uncommon that people who suffer a concussion keep going about their business for hours or days before finally feeling faint or confused. For a more common example, many of us do not realize that our brains are running low on fuel until we become light-headed and can't think straight anymore. Why is there such a disconnect between the state of our brains and our awareness of it?

For one thing, we cannot see our brains. But even more to the point is that we cannot *feel* them.

This is another major difference between the brain and the rest of our body. For all the chatter that goes on inside it, the one thing the brain is not very good at letting us in on is its own condition. This is not its fault, however. As opposed to everywhere else in the body, there are no *pain sensors* in the brain. As such, there is no way to feel "brain pain." If someone were to touch your brain, you wouldn't feel a thing. That's why surgeons are able to perform brain surgery on a patient even while the patient is still awake.

Most people confuse having a migraine or a headache as pain generated inside their brains. How often have we said something like "I had a migraine so bad I thought my head was going to split in half." Expressing it that way is more literal than we realize. It is in fact our heads that hurt, not our brains. This happens when the muscles in the neck and shoulders (not in the brain) stay semi-contracted for a period of time, for example, after spending several hours in front of a computer. Such tension can also radiate to the muscles in the face and scalp, activating the body's pain sensors located there and signaling discomfort. This is what most people confuse for brain pain. Next time you have a headache, take a good stretch.

What all this boils down to is that not only can't our brain cells be replaced, they can't even sound the alarm when there is a problem.

As a result, we remain oblivious to our brain's health. However, there are several things we can do to help our neurons stay strong and avoid unpleasant suprises down the line. Our ability to intervene and change the course of aging and disease is made possible by the remarkable abilities of the brain itself. Hundreds of scientific studies have shown that the human brain is quite the fighter. A healthy brain can take a lot of punches before being knocked out, or in biological terms, before exhausting itself to capacity. Then, and only then, do symptoms occur.

This is because the brain possesses its own reserve. Just as an auxiliary tank is to a car, the concept of "brain reserve" refers to the brain's capacity to perform in the face of ongoing attacks. Whether because of age, accident, or disease, if we let enough of these attacks stack up without care, our reserve will eventually be exhausted. Given this factor, a major clinical goal is to identify those people whose brains are silently fighting off diseases like Alzheimer's and initiate the preventative treatment necessary to save them from fully suffering their effects.

Hearing this, some of you might be hoping for a quick trip to the pharmacy, with a prescription in hand. Unfortunately, pharmacological treatments are limited. In the case of Alzheimer's, for once, current medications lessen or stabilize symptoms for a limited amount of time but cannot ultimately stop the damage Alzheimer's causes to brain cells. A new generation of disease-modifying drugs engineered to act as a vaccine is currently under development, but even the pharmaceutical companies can't guarantee that these drugs will be ready within the next decade. Meanwhile, clinical trials have yielded mostly disappointing results, confirming what everyone is loath to admit: treating a patient once clinical symptoms have emerged might very well be too late.

Pharmaceuticals are much more likely to work if applied during the earliest phase of the disease, well in advance of cognitive decline.

But for the time being, no such preventative treatment exists, or seems imminent.

This begs several urgent questions. Are we supposed to wait for still undeveloped drugs to emerge from a lab somewhere? Do we have time to wait? Could there be an equally effective alternative to a pharmacological approach? Above all, what can we do to make sure our brains stay healthy and active, preventing brain diseases from ever setting in?

NATURE OR NURTURE?

Recent research on Alzheimer's has shed light over the role we ourselves play in determining the future of our mental capacities. As you might know, Alzheimer's is often thought of as a nearly inevitable result of aging, bad genes, or both. It turns out that none of these alternatives is actually the case.

What most people don't know is that *less than 1 percent* of the population develops Alzheimer's because of a rare genetic mutation in their DNA. As we'll discuss in more detail later, the vast majority of patients do not carry any such mutation. So for the remaining 99 percent of us, the real risk is *not determined* by our genes.

This shouldn't come as too much of a surprise. Even diseases like cancer, obesity, diabetes, and cardiovascular disease in large part arise from the interplay of a multitude of genetic and lifestyle factors rather than from a single genetic mutation. Likewise, we need to recognize that the underlying causes of most forms of cognitive decline associated with brain aging, although sometimes partly genetic, are just as often linked to environmental and lifestyle factors like diet and exercise. So for the majority of us, whatever risk we do have has less to do with our genes and more to do with how we live our lives.

This is key in a very specific way. First of all, this indicates that our future (and that of our brains) is not beholden to our genes after all, but is largely dependent on the very choices we make. For instance, studies of twins provide fairly conclusive evidence that it is experience that molds our future, no matter what genes we possess. Research on identical twins is particularly enlightening in this respect. Studies of thousands of identical twins with the same exact DNA but who grew up in different home environments and led different lifestyles revealed that only 25 percent of their longevity was dependent on their genes. Therefore, it was their lifestyle rather than their genetic inheritance that had a much greater impact on whatever their risk ultimately was. These studies provide fairly conclusive evidence that it is experience that molds our future, no matter what genes we possess—and that it is those things within our control that can make all the difference, not only with regard to the quantity but to the quality of our years ahead.

In keeping with this, it was estimated that 70 percent of all cases of stroke, 80 percent of all cases of cardiovascular disease, and as high as 90 percent of all cases of type 2 diabetes in recent years were caused by nothing more than an unhealthy lifestyle. These diseases could have been prevented by simply paying more attention to dietary choices, weight modification, and physical activity. Importantly, there is recent evidence that addressing just a few of the risk factors for heart disease and diabetes could in turn prevent *over a third* of all Alzheimer's cases worldwide. These interventions ought to be even more effective in preventing or minimizing the less severe cognitive problems that naturally occur with age.

The truth is, we have more power than we realize. The power of our personal choices often remains untapped because of conventional Western medicine's tendency to treat symptoms with drugs or surgery before considering less risky and oftentimes more effective approaches instead—like eating better.

THE BRAIN-FOOD CONNECTION

For decades, the medical community has recommended dietary management as part of the therapeutic plan for many conditions such as diabetes, heart disease, high blood pressure, and high cholesterol. To date, no such recommendations exist for brain aging and dementia. In fact, many scientists and nonscientists alike are still reluctant to believe that our food choices might have something to do with the way our brains age or our risk of developing a brain disease.

In part, this is due to the fact that historically nutrition has been glossed over in medical schools, as well as in most post-grad mental health programs. It is only in recent years that nutrition was granted scientific-field status, and diet has been acknowledged as a legitimate means of protecting ourselves against brain diseases such as Alzheimer's. Little by little, scientists have come to appreciate the powerful connection between the foods we eat and our brain health. This very revelation has fostered a fast-growing body of evidence showing that we might very well be eating our way to dementia.

What many of us have only begun to grasp is that the actual health and quality of the foods we eat has dramatically diminished. Animals are routinely fed growth hormones, antibiotics, and genetically modified (GMO) feed, which we in turn ingest when we make a meal of them. Chicken and pigs are fed poisons like arsenic as a preservative. Conventionally raised produce is showered in pesticides and chemical fertilizers. In addition to being toxic and depleting our soil of nutrients, these treatments drive our produce to grow larger and plumper in appearance while disguising the fact that they possess an unprecedentedly diminished vitamin and mineral content. Additionally, chemically modified fats and refined sugar are routinely added to most foods. This is done not only to preserve the foods' shelf life but to deliberately increase our cravings for them, which in turn drives sales and profits.

What has gone unnoticed until now is the discovery of how, of all the organs in our body, the brain is the one most easily damaged by a poor diet. From its very architecture to its ability to perform, everything in the brain calls out for the proper food. Many of us are unaware that the only way for the brain to receive nourishment is through our diet. Day after day, the foods we eat are broken down into nutrients, taken up into the bloodstream, and carried to the brain to replenish its depleted storage, to activate cellular reactions, and, most important, to be incorporated into brain tissue. Proteins from meat and fish are broken down into amino acids, which, among other things, serve as the backbone of our brain cells. Vegetables, fruit, and whole grains provide important carbohydrates such as glucose, as well as the vitamins and minerals that energize the brain. Healthy fats from fish and nuts are broken down into omega-3 and omega-6 fatty acids that make our neurons flexible and responsive, all the while supporting our immune system and shielding the brain from damage. Our brains literally are what we eat.

FOOD AS MAGIC

Often scientists think of food as an assembly of calories and nutrients that have some specific interactions with human biology. If one follows that line of thinking, food found in Nature would be indistinguishable from food made in industrial factories. But Nature doesn't work like a factory.

Industrial food reflects roughly two hundred years of human innovation and research into nutrition, manufacturing, and optimization for human consumption. Natural foods, by contrast, reflect thousands of years of evolution and adaptation for life on the planet. When you put a blueberry in your mouth, for example, you benefit from all the effort (and thousands of years of trial and error) that the blueberry bush has placed not only into growing the berries, but also into protecting the future of the species, dormant in the seeds.

Instead of risking the seeds' survival on their own, the berries use their own defense system, which is composed of several chemical substances we humans refer to as *nutrients*—those very nutrients we work so hard to industrially decant into pills and capsules. Some of these nutrients are vitamins to prevent the seeds from getting spoiled. Others are minerals to give them strength. Others still are sugars to give them energy.

Additionally, plants produce a vast array of powerful compounds called *phytonutrients* (i.e., plant nutrients), such as the *anthocyanins* and *pterostilbene,* that have propelled blueberries into the news. Phytonutrients serve the important purpose of fighting oxidative stress and inflammation, thereby increasing the life of the seeds. They are also responsible for the berries' color, smell, and flavor. One reason those berries taste so good is that they are doing their best to attract the birds that eat them. That is because it is through the combination of the bird's digestion, flight, and excretion that the plant can spread its seeds far beyond its own territory, further safeguarding its survival on the planet.

In the end, by making berries, the plant sets out to ensure life. By eating berries, we receive all the benefits of that effort. There is a magic to this. Without anything like intention or thought, without advertising, laboratories, or a business plan, berries have made themselves nutritious and delicious. But far from being "supernatural," this is simply the magic of ordinary Nature.

In neuro-nutrition, there is an endless range of examples in which the nutritional whole (in its effects) is literally greater than the simple sum of its parts.

In spite of what our minds might tell us when presented with a brownie, what our brains actually crave is the multitude of nutrients present in natural, biologically active foods. When the right nutrients combine in the right way, the same magic that builds the healthy berries comes to build a healthy brain.

THE PROOF IS IN THE PUDDING

Could it really be this simple? Are you finding it all a little hard to believe?

I certainly did, once upon a time. But it was my very own research that convinced me otherwise. Let me show you how.

The figure below (Figure 1) shows the MRI scans of two healthy, dementia-free people following very different diets. Let's check out the differences between them.

To the left, you can see the brain of a fifty-two-year-old woman who's been on a Mediterranean-style diet most of her life. Not being a neuroscientist, you might not be able to recognize this at first glance, but her brain looks great. In fact, that's a picture of exactly how you want your brain to look when you are fifty-two. First of all, her brain takes up most of the space inside the skull (the white ribbon that sur-

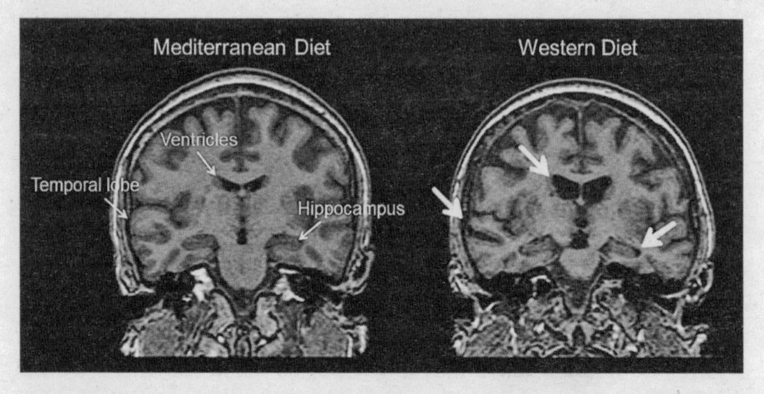

Figure 1. Comparing brains on different diets: Mediterranean diet vs. Western diet

rounds the brain in the figure). The *ventricles*, those little butterfly-shaped fissures in the middle of the brain, are small and compact. The *hippocampus* (the memory center of the brain) is well rounded and in close contact with the surrounding tissues.

In comparison, the scan on the right shows the brain of a slightly younger fifty-year-old woman who's been eating a Western-style diet for many years. This means fast foods, processed meats, dairy, refined sweets, and sodas. The arrows point to brain *atrophy*, or shrinkage, an indicator of neuronal loss. As the brain loses neurons, the space is replaced by fluids, which show up as black on an MRI. As you can see, there are more black areas present in the brain that has been fed a typical Western diet than in the brain that consumed a Mediterranean diet. The butterfly-shaped ventricles are larger in the brain to the right, which results from the brain shrinking. The hippocampus itself is surrounded by fluid (in black), as is the temporal lobe, another region directly involved in memory formation. These are all signs of accelerated aging and increased risk of future dementia.

Of course, not all people on the Mediterranean diet have perfectly healthy brains, and not all people who eat fast food have brains that are deteriorating. But on average, people who follow a Mediterranean diet seem to have overall fitter brains than those on less healthy diets, regardless of whether or not they carry genetic risk factors for dementia.

Findings such as these have led to a true and proper paradigm shift in medical practice, as an increasing number of experts now see diet as being as important to mental health as it is to physical health. In particular, there is mounting evidence that adopting a *brain-healthy diet* is key to maintaining optimal cognitive capacities well into old age, therefore delaying, or, even better, preventing the appearance of debilitating diseases like Alzheimer's. At the same time, eating well and leading a healthy lifestyle have the added benefits of reducing the risk and severity of other medical illnesses that also affect the brain, such as heart disease, diabetes, and various metabolic disorders.

In the end, science is teaching us that our brain health is highly dependent on the food choices we make. Though genetics can predispose us to many forms of disease, we should also give ourselves a little more credit when it comes to controlling the health of our brains (and bodies). What we all can and should do is be sure to take care of the brains we've been so gracefully given by nourishing them the best ways possible, which will naturally extend our chances of a longer, healthier life.

THREE STEPS TO NOURISHING YOUR BRAIN

The next chapters provide the information necessary to explore the promise of this alternate route to enhancing and protecting the health of our brains, while at the same time providing a guide for optimal cognitive fitness that applies far beyond any one disease or symptom. As we make our way toward this goal, my approach to neuro-nutrition provides three basic steps of care that you can easily incorporate into your daily life to enhance the health of your brain. The plan grows out of cutting-edge dietary concepts born of sound scientific research in neurology, biology, genetics, and nutritional medicine, along with the latest studies on food synergies and the microbiome.

Step 1: The first step is understanding which foods and nutrients your brain needs for optimal nourishment. Building a brain-healthy menu is the most important thing you can do to help your precious brain reach its full potential. Additionally, keeping your brain active—physically, intellectually, and socially—is also crucial to ensure optimal cognitive fitness.

Step 2: The second step is to improve and optimize your diet and overall lifestyle by following the general guidelines outlined in chapters

11 to 13. These are basic recommendations that anyone can adopt and develop, and in doing so, achieve far better brain health in both the short and long term.

Step 3: The third step is finding out where you are on the spectrum of your knowledge and practices as related to neuro-nutrition by taking the test included in chapter 14. This framework will help you reflect on where you stand in the process of nourishing your brain for the long-term. Beginner, already started, or further along—where on this trajectory are you? What can you do to move forward on the path to optimal cognitive fitness? Custom-tailored recommendations are provided for each of these three levels, presenting you with a full diagnostic workup for your own level in particular.

Via these three steps of discovering and embracing your customized blueprint, you will have integrated all that you've learned from this book to craft an optimal dietary and lifestyle plan that is good for *you* and you alone. Whether your goal is to boost your brainpower for the long haul, minimize memory lapses, or cut your risk of Alzheimer's, taking these simple brain-enhancing steps will help your brain be at its very best for the years to come.

2

Introducing the Human Brain, a Picky Eater

BONES AND BARRIERS

In order to fully understand how best to attend to the needs of our most complex organ, the brain, we first need to take a glimpse into its inner workings—what it's like, how it works, how it came to be the way it is. As we'll see in the following pages, the human brain is not only highly unique but also quite rebellious, working in accordance with its own rules—and, even, taste. Let's take a look, then, at this most incredible organ.

From the perspective of a neuroscientist, the brain can be described through a series of superlatives. It is first and foremost the most vulnerable organ of the body.

One would think that such an indispensable part of the body might have been made of indestructible materials. Instead, the human brain is quite soft. If you were to hold your brain in your hands, you would notice that it has the consistency of jelly. Thanks to the brain's containing a good amount of fat, its texture is delicate, rendering it vulnerable and easily damaged. Suffice it to say that it takes very little to harm the brain. So much so that Mother Nature considered it necessary to protect it with a skull made of layers of thick bones and to further swaddle it in several sheets of protective membranes called *meninges*.

This convenient built-in helmet with which we are all provided is fairly sturdy and resistant, enough so that an accidental smack on the head usually produces little more than an "ouch" or a grimace. But seemingly harmless accidents like these would result in major brain malfunction or even death were it not for the skull and meninges.

Inside the skull, the brain is immersed in a bath of colorless liquid called *cerebrospinal fluid* (the very same fluid that looked black on MRI), without which our ever-so-delicate brain couldn't even support its own weight. This fluid keeps the brain afloat and also provides cushioning from shock caused by sudden head movements, let alone hitting one's head on a hard surface. This fluid is also responsible for brain "clearance," the process by which the brain flushes toxins and waste away to keep itself clean and functional.

The skull, meninges, and cerebrospinal fluid are all necessary to provide structural protection and support to the brain—thereby granting it the status of "most protected organ" in the body. However, external insults are not the only danger to an organ that is so easily damaged. It also needs to be protected from within since many substances typically circulating in our bloodstream could potentially cause great harm to the brain. Nature saw to this by creating a special barrier that prevents these materials from entering the brain.

As further proof of its VIP status, the brain is the only organ known to be granted its own security system: a network of blood vessels called the *blood-brain barrier*. The blood-brain barrier is the brain's final layer of protection. It is made of a wall of flattened cells knit so tightly together that the barrier is fairly impermeable, preventing anything other than those elements recognized as safe and useful from entering the brain itself.

Think of the blood-brain barrier as a high-security government facility. Not unlike FBI headquarters, it sports monitored entrances and security guards at each gate. Some visitors are immediately recognized and allowed in. Others have to show their ID and pass through security equipment before being escorted inside. Still others are not allowed within even walking distance of the facility. This stratagem gives the brain total control over which substances are readily let in, which are allowed to cross only under supervision, and which dangers are to be kept at bay entirely.

Day after day, the blood-brain barrier protects the brain against infections and inflammation by restricting the passage of foreign, potentially harmful substances such as bacteria, toxins, and even some medications. At the same time, it exerts precise control over the entry of chemical messengers produced elsewhere inside the body that could interfere with brain activity, including some of our own hormones.

On the other hand, all sorts of substances that are necessary for the brain to function are allowed to pass through the barrier without hesitation. What is fascinating is that the vast majority of substances that are allowed to cross the blood-brain barrier are fragments of the air we breathe, the earth we walk on, and the foods that grow within it. It is quite moving to think that parts of the very planet we live on—its rivers, valleys, oceans, and skies—are routinely becoming part of our brains with every breath we take and every meal we consume.

Water first and foremost is a welcome guest. Water is always allowed to freely enter the brain, as are some gases like oxygen, which our cells need in order to breathe.

And what comes after water and air?

Nutrients.

Proteins, fats, and carbohydrates, as well as vitamins and minerals, keep our brains functioning, promote cellular activity, and prevent deficiencies. Over the millions of years during which the human brain evolved to be what it is today, highly specialized, nutrient-specific gates developed in the blood-brain barrier to allow our most important organ access to all the nutrients necessary for its growth and vitality.

FOOD FOR THOUGHT

The health of our brains, and our capacity to adapt and survive, is intrinsically dependent on our diet—and therefore our environment. In order to fully appreciate what an evolutionary advantage this represents, and how delicate the interaction between the brain inside us and the world outside is, we first need to look at how the human brain came to be.

Biomedical research is giving unprecedented attention to the importance of applying an evolutionary perspective to the origin and nature of our modern-day health problems. Over the last twenty years, research in evolutionary biology has shown that many of the key features that distinguish us from other primates go hand in hand with our distinctive nutritional needs. Such an approach has, however, been directed mostly at weight loss, physical fitness, and treatment of metabolic disorders such as obesity and diabetes—once again leaving the brain in the background.

Yet, when looking at our nutritional evolution as a species, it is the brain that has been impacted most. From prehistoric times to the present day, our brains have more than tripled in size, largely thanks to changes in our ancestors' diets and eating habits. The process of building these larger brains has been a slow and steady one that has taken place over some seven million years' time, alternating long periods of subtle increases in brain size with periods of dramatic growth spurts—which seem to have occurred in tandem with major dietary changes.

In the beginning, as far as brainpower was concerned, there was really nothing special about humans. For the first two-thirds of our history on the planet, the size of our ancestors' brains was within the range of those of some apes living today. For example, some of the earliest members of our family, the *Australopithecines*, had tiny brains approximately 400–500 cc in size.

Things remained relatively unchanged for millions of years until *Homo erectus* made his debut about 1.8 million years ago, sporting a notable 1000 cc brain. The human fossil record indicates that this was the first substantial burst of evolutionary change in brain size. The next leap didn't take as long. Brain expansion accelerated significantly in the last 500,000 years, gifting *Homo sapiens* and their relatives the Neanderthals with brains that were roughly the same size as our modern-day brains (1300–1500 cc).

The end result is that we have an enormous brain for an animal of our size. In comparison, chimpanzees, our closest living relatives, are very similar to us in body size but have brains that are only one-third the size of ours. Most of this size difference reflects the expansion of the parts of the human brain specializing in sophisticated cognitive functions such as language, self-awareness, and problem solving, which also afforded us the development of tool making, symbolic thinking, and socialization—all skills that make us human, while at the same time allowing us to take better care of ourselves.

But brain expansion came at a very high cost for our species. Besides the fact that it's not easy to carry around, a jumbo brain requires proportionally more energy and therefore considerably more calories and nutrients. Compared to other primates and mammals of our size, humans must allocate a much larger share of their daily energy budget to feeding their brains.

In order to meet the growing brain's energy demands, our ancestors had to seek out a diet denser in energy and fat, which must have been very expensive in terms of the time and resources necessary to supply these. Regardless, the ability to maintain this diet was vital for our ancestors. A large body of scientific evidence suggests a direct relationship between access to food and brain size, where even small differences in nutritional quality had a large effect on survival and reproductive fitness. Perhaps Nature decided that this trade-off was worth the effort for the sake of our progress. After all, it is this dramatic brain expansion that first stirred humans to dream of early cave paintings, later realize the first moon landing, and, most recently, invent and expand the online universe we call the Internet, ultimately setting our species apart from all other animals.

THE "PALEO BRAIN" DIET

The role of diet and nutrition in shaping humankind has intrigued scientists and gripped the popular imagination for some time. As hard as it is to study the diet of *Homo sapiens* and their extinct relatives who lived millions of years ago, paleontologists have been able to reconstruct the evolution of our predecessors' diets with remarkable accuracy and detail.

Geography provided their first clue. Africa has long been agreed upon as the cradle of humanity. But the Africa of our ancestors was not the hot and parched Saharan desert we think of now. Instead, the re-created environment of early man suggests a mosaic of extensive grassy woodlands, gallery forests, and wetlands with fluvial floodplains, producing lakes and swampy, vegetated areas. It is from this lush, verdant niche that the first hominids emerged millions of years ago, with their small heads and wobbling gait. These early humans had an appetite for fruits and leaves, and consumed a diet similar to that of the modern ape. Grasses, seeds and sedges, fruits, roots, bulbs, tubers—even tree bark—were the most likely sources of nutrition for our ancestors, whose massive jaws, robust faces, and large molars lent themselves so well to the slow, thorough chewing necessary to assimilate these foods.

But the low-calorie content of this diet makes it an unlikely candidate for providing the energy necessary to promote significant brain growth. The major increase in brain size heralding the appearance of *Homo erectus* would never have happened had our ancestors been satisfied with eating stems and flowers. As their brains were slowly getting bigger and stronger, they were also growing *hungrier.*

So what happened to spur such dramatic brain expansion?

For many years it was believed that our predecessors took in the extra energy needed to fuel their expanding brains by forgoing the low-calorie plant diet they were used to in favor of eating calorie-dense meat. This would make perfect sense had that meat been easier to come by. But early humans lacked the ability to hunt game themselves. They were relatively small-bodied and their primitive ape-like features were not made for chasing big animals. Further, our ancestors' brains were fairly small, and their skills just as limited. It wouldn't be until a few million years later that the *Homo erectus* would develop a marked increase in both brain *and* body size, thereby developing the limb proportions and posture necessary to efficiently run after prey.

The age-old paradigm of "man the hunter" has suffered a big blow in recent years, as research revealed that our role as hunters has been exaggerated. Contrary to what popular imagery suggests, and as much as meat might have been a highly valued food, it remained a rare and dangerous one to actually obtain.

If it wasn't meat that did the trick, then what was it?

As it turns out, it was none other than fish.

There is abundant paleo-environmental and fossil evidence that early humans lived in proximity of water. Fresh drinking water is indeed the single most important resource for the human body. Throughout history, we've done our best to set up civilization within easy reach of it. An added bonus to living next to rivers and lakes is that other animals also lived by them—and most important, *in* them. The East African Rift Valley, with its extended fluvial networks and luxuriant vegetation, has probably been the exceptional ecological niche that spurred the brain's expansion by providing energy-dense "brain food." Back then, the shorelines were a lavish source of shallow and low-water aquatic species, supplying snails, crabs, mollusks, sea urchins, and small fish, as well as fish roe, spawn, amphibians, and reptiles. When the day's catch went awry, insects and worms were likely plentiful year-round and bird eggs available on a seasonal basis. Moreover, these lands also had an almost limitless abundance of plants, fruits, vegetables, and weeds.

What makes this habitat particularly suitable for brain development is that these foods required little skill to fetch and consume, favoring the less coordinated skills of a smaller brain, while at the same time containing the perfect nutritional content needed to promote brain growth. Fish and shellfish are excellent sources of polyunsaturated fat, which is code for the omega-3 composition that makes headlines today—the very fat that our brains are in large part made of. They also contain a bounty of proteins, vitamins, and minerals that are essential for brain function. Vegetables and fruits provide even more vitamins and minerals along with brain-friendly sugars, while eggs contain precious nutrients such as *choline*—a substance used by the brain to memorize information and learn from experience. We'll talk a lot more about these foods and nutrients in the next chapters.

Now back to where we left off. There is evidence that, besides feasting on easy prey like eggs and shellfish, early humans participated in "confrontational scavenging," allowing other animal hunters to make the kill before chasing them away to take the carcass. Evidently, their brains had already evolved enough to provide them with the necessary skills to outsmart their competitors. This not-so-gentlemanly behavior further increased access to land and sea animal protein, including birds, turtles, amphibians, even crocodiles. As unpalatable as these animals might seem to our modern taste buds, they definitely provided an additional source of nourishment for our greedy, ever-expanding brain.

Eating better made our ancestors smarter, and being smarter now allowed them to feed themselves more efficiently. Little by little, as the brain grew larger, man grew taller. At the same time, eye-hand coordination improved and planning skills became more sophisticated. In the process, our ancestors learned to stand, to walk, and, finally, to run, which along with newly developed hunting techniques enabled them to start catching birds and small mammals, and eventually hunt for bigger and faster (not to mention fresher) game.

This higher-quality diet further increased our ancestors' fat consumption and energy budget, and proved crucial to the rapid brain evolution of *Homo erectus*.

Animal foods like fish and meat proved helpful in many ways. First, they are both excellent sources of fat. Strong bones require minerals, and fat must be present for those minerals to be absorbed. Fat also helps regulate body temperature, hormone production, and blood pressure. But even more important, fat acts as the body's reserve tank of energy. When sustenance depends on foraging for fruits and vegetables with only the occasional opportunity to hunt wild game, being able to store fat calories was often the difference between life and death. Just like bears who spend their summer and fall eating in preparation for their long, deep sleep, our bodies learned to do the same. This new ability to store fat for future fuel represented an incredible evolutionary advantage for our hunter-gatherer ancestors, especially when confronted with periods of food shortage. It is difficult to imagine scarcity while surrounded by supermarkets and online grocery stores 24/7. But prior to modern times, man was beholden to Nature's relentless cycles. There were seasons of relative abundance and seasons of hardship. When meat and fruit were scarce, which was often, our ancestors relied on whatever the land provided, which was oftentimes not much more than plants, nuts and seeds, tubers, wild grains . . . and bugs.

But even in times of abundance, contrary to what many people believe, it wasn't "man the hunter"—or more specifically "man the fisherman"—who procured the most dietary protein and fat. In reality, the majority of energy-dense food in our ancestors' diet didn't come nearly as much from men hunting or fishing as much as it did from women gathering. Studies show that even today over 65 percent of the food in worldwide hunter-gatherer communities is provided by plant-based sources, while no more than 25 to 35 percent is derived from hunting game.

Additionally, while a basic tenet of the so-called Paleo diet is that early humans did not eat grains, new evidence has emerged that people enjoyed their carbs well before the Paleolithic era (which is long enough to evolve the capacity to digest them). Several research teams have documented how ancient grains like oats and wild wheat were a recurrent feature on our ancestors' menu as early as 3.5 million years ago. Basically, people ate what they could get their hands on. Eating was surviving.

Throughout this long search for better and richer foods, we not only succeeded in developing larger brains and bodies. We also became more ingenious as to how to access and assimilate those foods.

MAN TURNS CHEF

Another major turning point in brain evolution occurred when man mastered the use of fire. Although our ancestors discovered fire almost 3 million years ago, it took quite some time to govern this tricky element. It could very well be the development of habitual cooking, supported by the building of stone hearths and clay cookware, that fueled the brain's latest growth spurt of half a million years ago.

Pounding and heating food, most likely in the form of roasting meat and vegetables, render the nutrients easier to digest and absorb. Cooking produces soft, energy-rich foods, which means less time spent chewing and digesting, while preserving calorie content. This left more time and energy to focus on other activities—like growing bigger brains. It is even possible that our brains intuitively engineered all of this in the first place. After all, cooking is among our uniquely human abilities.

In addition to its effects on the brain, eating cooked foods also contributed to reshaping the human body. Earlier humans needed big teeth and jaws to break down bulky plant fiber, not to mention bigger gastrointestinal organs to further facilitate absorption of nutrients. But thanks to the increased access to animal foods and their own cooking skills, the *Homo sapiens* no longer needed such cumbersome apparatus. Teeth, jaws, and guts got progressively smaller while heads got bigger, effectively swapping guts for brains. As a result, our gastrointestinal tract is characterized by a larger stomach and a shorter colon than that of purely herbivorous animals, though not as short as carnivores, making humans able to extract plenty of nutrients from both animal and plant food.

Flash back to ten thousand years ago, when the human diet took another major turn thanks to the development of agriculture and farming techniques. The ability to grow crops as well as to raise livestock provided a steady, ongoing access to food, giving us an unprecedented nutritional advantage. This new, plentiful, and reliable food supply of grains such as barley, wheat, corn, and rice, as well as eggs, milk, and domesticated meat, led to more regular meals and at the same time afforded us a more sedentary lifestyle. This allowed the population to grow faster than ever before. However, though we are generally better fed, we have also become relatively under-muscled and plumper when compared with other primates. Additionally, the changes in body fat composition that helped to offset the high-energy demands of our brains started making our race more prone to diseases of affluence like obesity and diabetes, which can instead affect the brain. As a result, humans' propensity to store fat is not much of an advantage when one is surrounded by fattening foods.

Since the agricultural revolution, and even more notably the Industrial Revolution of the late 1700s and early 1800s, which saw the mechanization of agriculture along with the advent of railroads and steam ships, humans have radically changed their lifestyle and dietary habits . . . though not for the better.

WE ARE WHAT WE . . . ATE

When we review its contents today, our ancestors' diet couldn't be more different from ours with regard to both quality and quantity. The plummet in quality is especially stunning. Where our ancestors ate mostly vegetables, fruits, nuts, and seeds, many Americans today barely touch these foods. When they do, it is rarely in their natural state but rather canned, frozen, juiced, or otherwise chemically processed, and usually eaten as a side dish or snack rather than a main course. Things are not much different in other parts of the industrialized world.

Whereas carbohydrates used to come from fresh, seasonal fruits and vegetables (and sometimes honey), now we access them from processed grains, cereal, and most commonly straight-up refined-sugar products. In addition, the wild animals that used to roam the savanna contained more protein and a higher amount of brain-friendly fat than the unfortunate, domesticated chickens and cows that have become staple foods in our modern diets. Even today it is clear that wild-caught fish possess a much healthier omega-3 fat profile than farm-raised fish, and not nearly as many toxins and pollutants. Worse still, our fat consumption is relegated to processed baked goods, dairy products, margarines, and adulterated butters, which include precisely the opposite proportions of unhealthy to healthy fat.

As a result, the modern Western diet is a disastrous combination of refined grains, processed meats, and dairy products that have been stripped of any nutritional value. At the same time, we've all but eliminated fresh, organic fruits and vegetables. And how much of our weekly diets include wild-caught fish?

Throughout these massive dietary revolutions, one thing has become painfully clear: our brains haven't managed to keep up.

There is growing evidence that our DNA hasn't had the proper time to successfully adapt to the dramatic lifestyle and nutritional changes that have occurred so rapidly and so recently during our evolutionary time scale. This very fact has major implications when we look at our present-day dietary needs.

Humans have been on Earth for over 5 million years and have lived as hunter-gatherers for 99 percent of this period of time. While diets can change in the blink of an eye, our genetic makeup is not so flexible. Most genetic variations that make us who we are today were already present when *Homo sapiens* emerged some 500,000 years ago. That also holds true for many, if not most, of the disease-causing genes known to scientists thus far. As a result, our brains are just not genetically prepared to consume the modern diet.

In a day and age where we have lost the instinct we once had to eat right for our brains, and are often misguided by the media and the outdated education we receive, we have to take it upon ourselves to relearn what's truly good for us.

According to some popular current ideas, we could begin by looking at the foods and nutrients that made brain evolution possible in the first place. However, it's not quite as simple as eating and exercising as our ancestors did—just as it's not possible to bring back their environment and living conditions. Rather, strange as it sounds, we need to listen to our brains to get a better understanding of what's needed to reach optimum health and mental sharpness.

3

The Water of Life

YOU'RE IN—OR YOU'RE OUT

When I first became interested in neuro-nutrition, I realized that there was a lot of confusion over which foods and nutrients were good for the brain and which were harmful. Depending on the day and whom you listened to, you would hear that eggs are good for you one day and bad the next, that sodium is responsible for high blood pressure . . . until it's not, while carbs and fats take turns playing hero and villain.

Personally, I find that much confusion is due to the fact that few health-care professionals know how the brain actually works, and fewer yet have the advantage of having studied brain chemistry.

What most people don't realize is that the nutritional requirements of the brain are substantially different from those of the other organs of the body. As we have begun to see in the previous chapter, the human brain is a very peculiar organ, working in accordance with its own rules and preferences. We are now going to see how the brain is highly unique also in terms of its diet. As hungry as it might be, the brain is at the same time a very picky eater. In comparison with the rest of the body, which figured out a way to process most nutrients to its advantage, our brains are very strict and highly selective when it comes to food.

If we were to compare the human body to the world's food trading system, we could say that the brain asserts austere international trade regulations in comparison to other organs. In the real world, if a country can provide enough food on its own, it can limit the importing of those foods from elsewhere. When you eat fresh blueberries with your oatmeal on a winter day, it's because those blueberries were imported from South America. But if you are adding milk to that same oatmeal, it most likely comes from a U.S. farm.

The brain is just as conservative when it comes to importing food. Whatever the brain can make locally is made locally. Yes, you read correctly. The brain has the capacity to supply its own nutrition. Not all of it, mind you, but some of it. Everything else must be obtained from the food we eat.

"Everything else" means all the nutrients that the brain needs but cannot make itself (or make enough of to meet its needs)—which I'll refer to as *brain-essential* nutrients. How do we know which nutrients are brain-essential and which aren't? For starters, brain-essential nutrients have the distinct honor of being among the few substances that are able to cross the blood-brain barrier so that they can successfully arrive at our brains in the first place. This is when studying brain chemistry comes in particularly handy.

The human brain requires more than forty-five nutrients to be at its best, and the ways these nutrients are used are as different as the molecules, cells, and tissues they help to create. Nutrients are usually divided into five major groups, representing the basic components of our food: proteins, carbohydrates, fats, vitamins, and minerals.

Now for a very distinctive difference between our bodies and brains. On average, our bodies are made of a fair amount of water (60 percent), followed in prevalence by proteins (20 percent), fats (15 percent), carbohydrates (2 percent), and some vitamins and minerals. These proportions shift in the brain, since water content is even more prevalent there than in the rest of the body. In fact, the brain is made of almost 80 percent water. That's quite a lot of water for such an active organ. Fats (i.e., lipids) come in second (approximately 11 percent), followed by proteins (8 percent), vitamins and minerals (3 percent), and a pinch of carbs.

We'll discover even more differences between body and brain as we start to look into each of these nutrients in more detail. In the meantime, let's let the brain lead the way as we begin our exploration of neuro-nutrition, starting with the most prevalent brain nutrient of all.

THE THIRSTY BRAIN

If you don't think of water as being nutritious, think again. Many scientists believe that life on Earth itself was made possible by the presence of water on the planet, and that the very first living creatures were born in the depths of the oceans of 4 billion years ago.

Water is undeniably vital to human life and, as it turns out, also to our intelligence. Besides constituting most of its weight, water is involved in every chemical reaction occurring in the brain. In fact, brain cells require a delicate balance of water and other elements such as minerals and salts to work efficiently at all. These *electrolytes* (minerals and salts that help you stay hydrated, such as chloride, fluoride, magnesium, potassium, and sodium) flow in and out of your brain with every sip of water you drink. Further, water is indispensable for *energy production*—that's because it carries *oxygen*, which is needed for your working cells to breathe and burn sugar to produce energy. Water also plays a structural role, filling in the spaces between brain cells, and also helps to form proteins, absorb nutrients, and eliminate waste products.

As further proof of its importance, we can last weeks without food but only days without water. The body can't store water, so we need a fresh supply every day to make up for the loss that occurs from the functioning of our lungs, skin, urine, and stools. When water supply is too low, we become dehydrated. Dehydration occurs when we use or lose more water than we take in, and the body doesn't have enough fluids to carry out its normal functions. This could be surprisingly dangerous, especially to our delicate brains. Dehydration disrupts energy processes and causes loss of electrolytes, a loss to which the brain is particularly sensitive. It has been estimated that as little as a 3 to 4 percent decrease in water intake will almost immediately affect the brain's fluid balance, causing a number of issues like fatigue, brain fog, reduced energy, headaches, and mood swings, for starters. If 3 to 4 percent sounds like a small degree of water loss, that's because it is. You could easily reach that level of dehydration just by going about an average day that included moderate exercise and neglecting to drink water throughout.

What's worse, most people don't drink nearly enough water to start with. According to a recent study by the Centers for Disease Control and Prevention, 43 percent of adult Americans report drinking less than 4 cups of water a day, including 36 percent who drink 1 to 3 cups, and 7 percent who drink none.

This is particularly disconcerting, as dehydration was shown to accelerate the brain shrinkage that occurs with aging and dementia. MRI studies show that when we are dehydrated, several parts of the brain appear to get thinner and lose volume. Clearly this is a much more pressing issue than many might have thought. The good news is that the effects of dehydration can be fully reversed in a matter of days by simply drinking more water.

There is some debate over how much water we really need to drink. On average, the recommended amount is eight 8-ounce glasses a day. If you need more of an incentive, this number is backed by research showing that drinking 8 to 10 cups per day can boost your brain's performance by almost 30 percent. Researchers in the U.K. ran an experiment to test the potential effects of water on cognitive performance and mood. They had several participants complete a series of mental tests after eating a cereal bar. Some participants consumed just the cereal bar alone. The rest were also given water to drink. Those who drank around 3 cups of water just before completing the tests showed significantly faster reaction times compared with those who did not drink any water.

Think about the advantages of thinking faster. Drinking a glass of water first thing in the morning can increase your ability to shower, have breakfast, get dressed, and be out the door right on time to catch the subway. Conversely, your ability to calculate how much time this all takes is compromised when your fluids are low, and you end up hitting the snooze button for another half hour (and missing the train).

Overall, the prescription is not difficult to follow: make a point of drinking eight 8-ounce glasses of water a day, or close to 2 liters (2 quarts). Then adjust the amount based on your own particular needs. Depending upon your age, environment, and activity level, you might need more. If you live in a warm climate, you'll need to drink more water than someone who lives in Alaska. If you are a professional athlete, you'll need to drink more water (and electrolytes) than someone sedentary. Plus, we all need more water as we get older. For reasons yet unclear, aging alters thirst and drinking responses, making older people more vulnerable to fluid imbalance in their brains, which might very well contribute to cognitive decline and neurological disorders like Alzheimer's.

As a neuro-nutritionist and brain health advocate, I'm convinced of the value of drinking enough water. Of all the tricks I've learned for keeping my mind sharp, staying hydrated might be the one I follow most religiously, starting with a glass of water first thing in the morning (which is essential after a night without any fluid intake) and ending the day with a cup of herbal tea.

WHAT DO YOU MEAN BY "WATER"?

When doctors prescribe water as a key part of one's health regime, what precisely do they mean? Are they talking about tap water or filtered water? Does seltzer count? And herbal tea—isn't that practically water? Don't we get some benefits from the water content in juices or caffeinated beverages, too? There are, after all, many people who drink very little water and manage to survive. Let's look a little more closely at what's what.

Sure enough, according to the U.S. Department of Agriculture (USDA), carbonated soft drinks are the most-consumed beverages in the United States, along with purified bottled water and beer. Milk, coffee, fruit juice, and sports beverages come in a close second, as well as iced tea. Wine and distilled spirits are also present on the list but in smaller amounts per capita. Is this enough "water"?

From a purely scientific perspective, anything that increases water content in the body counts as a *fluid*. That said, there is quite a marked difference between drinking a glass of spring water and having a cup of coffee. While the extent of this reaction varies with each person, many of us are sensitive to the caffeine contained in our coffee. When you are consuming water in coffee or black tea, the caffeine is actively dehydrating you as you drink it, rendering its water content decidedly less than effective. Soft drinks, on the other hand, contain water but not nearly as much as they contain refined sugar. Even milk could be tricky. In its purest form, milk contains water as well as several important nutrients. However, in many industrialized countries, milk is one of the most highly processed products on the market, and widely overchugged at that.

The bottom line is, replacing water with beverages that contain unwanted fat, hard-core sugars, artificial sweeteners, preservatives, and coloring, and that at the same time promote dehydration and weight gain, clearly defeats our purpose.

The longevity and well-being of both your brain and your body are critically dependent upon your consumption of what we call *hard water.* This refers to plain water that is high in minerals like calcium and magnesium.

Hard water isn't hard to find. Being a fan of all things natural, my preference is to drink spring water. Whether from an unpolluted underground spring in France or from an artesian well in Fiji, spring water comes from rain and snow. After being collected in natural basins, it is slowly filtered through layers of rock, where it picks up all sorts of valuable minerals, salts, and sulfur compounds.

Sparkling water that comes from a natural spring also contains various healthful minerals. In this case, the carbonation isn't added by the bottler but from the spring itself. That means that the bubbles in these bottles are completely natural. The only downside is that it can be quite expensive.

Seltzer and club soda, instead, are just plain waters that have been artificially carbonated. Seltzer contains no sodium salts (and therefore won't help you stay hydrated), while club soda usually contains mineral-like ingredients that are added by the bottler to enhance the flavor of the water. As we'll discuss in more depth in the next chapters, lab-produced nutrients aren't nearly as effective as natural nutrients at improving health, metabolism, or resilience against disease. My personal recommendation is to stick to natural hard water as much as you can.

This reminds me of my first trip to the supermarket in New York City, soon after my arrival in the United States. Having found the tap water lacking, I went in search of bottled water. But that was easier said than done. First, to my surprise, I was directed to the refrigerated aisles. As I later learned, bottled water can be found on the shelves as well as refrigerated, but most Americans prefer to drink their water cold—even better, with a few ice cubes thrown in. It never occurred to me to look for bottled water in a fridge. Most Italians drink their water at room temperature as it's easier on the stomach, so our bottled water is rarely sold cold. To this day, I abide by that principle. Finally, once I located the fridge, I was so overwhelmed by such a plethora of seltzers, sodas, and sports and energy drinks, along with all the juices, flavored milks, and smoothies, that the water options paled in comparison. The final blow was that once I located the water selection, most bottles sported a "purified" label that baffled me.

As it turns out, purified water is currently the most consumed water in the United States. Unfortunately, although the word *purified* usually implies a good thing, in this case it refers to water that has been stripped of its chemical content—thus wiping out all its precious minerals in the process. Although purified water might be safer to drink than tap water, it has been rendered nutritionally void. As a result, it is entirely incapable of hydrating you or your thirsty brain. If you drink purified water because you're concerned about impurities, you should take mineral supplements along with it.

Even better, consider filtering the water yourself. Personally, since nutrient-rich bottled water can be expensive, I invested in a high-quality faucet filter to filter tap water at home. Tap water is needed for other things besides drinking and cooking, such as washing and cleaning, so it's important to ensure that it is as clean as possible. The goal is for your filter to get rid of harmful chemicals like asbestos, chlorine, lead, benzene, and trichloroethylene—all of which can be found in tap water. Filters can take out what's harmful in your water while leaving its good mineral content intact. There are many available filtration options, so if you are interested in learning more about them, speak to a specialist. I can attest to the fact that installing a water filter is easier and more affordable than you might think, and it is a great first step in making sure that you and your family have easy access to plenty of drinkable, nutritious water on a daily basis.

Contrary to what many people believe, hard water is also better than sports and energy drinks for rehydrating after a workout. The vast majority of these beverages are high in sugar and sodium, in addition to being chockful of manufactured minerals and salts, and therefore are not good for you. If you feel that water alone is not enough, my recommendation is to try coconut water. Coconut water is Nature's thirst quencher, being low in sugar while still providing you with the potassium you're looking for. Most coconut water contains up to 300 mg of potassium and as little as 5 mg of natural sugar per glass, unlike your favorite sports drink that contains only half the potassium you need and five times the amount of processed sugar you don't want. Be sure to go for unsweetened, organic coconut water to replenish the fluids your body loses during a workout or even just during the course of the day.

In addition to drinking plenty of water (and coconut water when I need it), I rely on a favorite trick: aloe vera juice. Aloe juice is Nature's own first aid—naturally anti-bacterial, anti-viral, and anti-fungal, this healthful juice contains about 99 percent water and over two hundred active components from vitamins and minerals to amino acids, enzymes, and even fatty acids. Thanks to its well-established properties, drinking aloe juice is a great way to soothe and hydrate all your organs, reduce inflammation, *and* get the brain ready to spring into action, from the inside out.

Also keep in mind that drinking is not the only way to maintain water balance. According to the recommendations of the Institute of Medicine of the National Academies (the advisors to the nation on food and health), we could actually eat, rather than drink, up to 20 percent of our daily water intake. This means that one to two glasses of our daily fluids can be obtained from water-rich foods rather than water itself. To do so, focus on fruits and vegetables. These fluid-filled foods will support your staying hydrated while also providing healthful nutrients, not to mention natural sugars that the body can actually use to its advantage. Take a look at Table 1 for a list of fruits and veggies boasting the highest water content. Among vegetables, cucumbers and lettuce top the list with a whopping 96 percent. Next are zucchini, radishes, and celery, followed by tomatoes, eggplant, and cruciferous vegetables like broccoli, peppers, and spinach. Perhaps it comes as no surprise that watermelon has the highest water content of all fruit (93 percent water per volume), followed by strawberries, grapefruit, and cantaloupe. By comparison, a banana provides a relatively low 74 percent.

In conclusion, simply drinking more water and eating water-rich foods could be one of the healthiest changes you make to bring about better health as well as greater cognitive power in your life.

Fruits	Water content	Vegetables	Water content
Watermelon	93%	Cucumber	96%
Strawberries	92%	Lettuce	96%
Grapefruit	91%	Zucchini	95%
Cantaloupe	90%	Radishes	95%
Peaches	88%	Celery	95%

Table 1. Top five fruits and vegetables with high water content

4

The Skinny on Brain Fat

WHAT IS BRAIN FAT?

Throughout the book so far, there have been hints about how important some nutrients are to the health of our brains. There have also been hints about how misrepresented some other nutrients are in relation to brain health. Fat is among the latter.

Most people are aware that the human brain is rich in fat, which accounts for about 11 percent of the brain's weight. However, I bet you expected a much higher percentage of fat in the brain than 11 percent. If you did, it could be because you've heard somewhere that the brain is *made up of fat*, usually 60 percent or higher. If you haven't heard that, just google "brain fat." As of March 2017, there were 127,000,000 results that confirmed precisely that. There might be even more by the time you read these pages.

I was always puzzled by these statistics and had never been able to find any scientific study that confirmed such a high prevalence of fat in the brain relative to other nutrients. From a purely biological perspective, the mismatch is in part due to whether you include water in the calculations. Basically, there are two ways to evaluate the brain's composition: you could estimate it including its water content or in terms of the brain's "dry weight," which excludes water. Including water lessens the fat content, while excluding it increases the fat content. If you include water in the calculation, fat makes up about 11 percent of the brain's composition. If you take water out of the equation, fat accounts for less than half the brain's weight. Incidentally, there isn't much more fat than protein, which made the Internet posts even more puzzling to me. The problem is that many people, including several journalists and even a few doctors, believe that eating fat is good for the brain—the argument being that since the brain is "made of fat," like cholesterol, eating that fat must be good for you. On the other hand, the vast majority of scientists will tell you exactly the opposite. Unfortunately, scientists rarely appear on the evening news and their voices remain confined to academic journals nobody has access to except other scientists, which further complicates an already complex picture. This lack of clarity has become problematic from a medical perspective, and brings us to a discussion of good versus bad fats, a discussion that features prominently in much research and seems to be making headlines of late.

While the prevailing wisdom of the past several decades suggested that low-fat, high-carb diets were the most beneficial for human health, society is now in full "fat redemption" mode. This trend is well exemplified by the June 2014 *Time* magazine's thought-provoking cover story "Ending the War on Fat"—an attention-grabbing headline juxtaposed over an image of a generous dollop of butter.

It turns out that Americans are eating more full-fat foods than in previous decades, to the extent that experts expect the pro-fat trend to continue over the next fifteen years. Butter sales in the United States went up 20 percent in 2015. The number of people buying whole milk has gone up 11 percent, while skim milk sales plummeted 14 percent. In addition, everybody is ditching carbs for fats. Part of this dramatic shift in dietary habits is due to science's change of heart in its position on fat and our health. Recent research has shown that high-carb diets can promote weight gain as much as high-fat diets, and that eating cholesterol has only a modest influence on the level of cholesterol in the blood and, therefore, a lower impact on the risk of heart disease than previously thought.

While dietary trends come and go, a common misconception seems to have remained unaltered regarding brain health. As I learned from my research participants, the public is persuaded that eating fat is necessary to make sure our brains work correctly. But when patients are asked what a high-fat diet should consist of, it turns out that many, if not all, are suggesting a sizzling slice of bacon, a nice piece of cheese, perhaps a scoop of sugar-free whipped cream. Are these foods really good for your brain?

The answer is *no*.

It's time to bring some clarity to this controversial issue. The human brain contains fat—that much is true. But it might not be the kind of fat you're thinking of. The truth is, *fat isn't just fat*. There are many different types of fats (or *lipids*). Some fats have become household names, like cholesterol. Cholesterol is one of the fats we are all familiar with thanks to our doctors' routinely measuring them in our blood work.

But there are many more and likely unfamiliar fats inside our brains, like *phospholipids* and *sphingolipids*. If you are thinking "What and *what*?" you are not alone. Most people have never heard of these terms. The truth is that these largely neglected fats, rather than the familiar ones, account for as much as 70 percent of all the fat found in the brain. There are many inconsistencies like this one in "popular neuroscience" that have generated much confusion about which nutrients are really important for neurological health. There will be many more to come—*and* to correct.

NOT ALL FAT MAKES YOU FAT

Another common misconception is that the brain needs fat for energy. That's because we all intuitively know that our *bodies* will use the fat contained in fattening foods to give us energy. Instead, despite containing quite a bit of fat, the brain actually lacks the ability to burn that fat to produce energy for itself.

Reading this might lead to raised eyebrows and puzzled faces. Don't we burn fat to make energy? Didn't you say our ancestors really started to thrive once their bodies learned to store fat as an energy source? What does the brain do with all that fat if it can't use it? This "brain-fat paradox" has been giving a headache to many nutritionists and dieters alike.

The reality is that when it comes to fuel, the brain doesn't touch any of this fat. It simply can't.

What most people don't realize is that there are two types of fat in the human body: storage fat and structural fat.

When we talk about "fat," we are usually talking about storage fat (also called *adipose fat* or *white fat*), the fat that our body uses to store energy. Storage fat is the soft, squishy, visible fat you can actually see on your body. What about that muffin top people talk about losing? *That's* storage fat.

Storage fat is in large part derived from our diet. When we eat foods that contain fat, such as meats, cheeses, or candy bars, their fat is broken down into smaller units like saturated fat. This process itself generates heat and energy. These fats are then reassembled into *triglycerides*, the molecules that your doctor routinely checks in your blood work in tandem with cholesterol. Triglycerides are then stored away in adipose tissue for future fueling. This makes storage fat the largest energy reservoir we possess as human beings, acting as a convenient, portable battery pack.

A typical 160-pound (~70-kg) man has almost 27 pounds (12 kg) of adipose tissue fat, equaling roughly 100,000 calories' worth of energy. If he were starving, this reserve would supply him with fuel for nearly two months' time. So as far as your body is concerned, having some "meat on your bones" is a very good idea.

On the other hand, the brain does not contain that sort of fat at all. In fact, there is no storage fat in the brain.

Whatever fat is found in our brains is a totally different kind of fat called *structural* fat. The rest of our bodies also contain plenty of this fat. Structural fat is just as essential to life as storage fat is to fueling us, but the way our body uses these two fats couldn't be more different. For starters, structural fat is not utilized as energy. Second, it isn't the type that floats around in our bloodstream or clogs our arteries. Instead, as the name indicates, it is used to structure our cells and as a sort of "technical" support. For example, as far as the brain is concerned, brain cells are wrapped in a fatty sheath called *myelin*, which provides a clever insulation for electrical impulses traveling to and from the brain. Whatever cholesterol the brain contains is found mostly on this myelin sheath, acting as an insulator. Neurons are also enclosed by delicate fatty membranes that not only protect them from external insults but also allow signals and nutrients to flow in and out of the cell. These membranes are made of other kinds of fat, like the famous omega-3s and the largely overlooked phospholipids. There are many more examples of how brain fat is used exclusively for structure rather than for energy.

A likely explanation for this is that if the brain had been made of combustible or "burnable" fat, it might start eating itself in case of starvation—which would certainly put us at quite an evolutionary disadvantage.

In the end, the only fats we actually need to feed our brains are the building blocks that support its structural health. If you're wondering which foods contain these fats, here's another surprise. Just because these fats are present inside your brain doesn't mean you have to eat them.

FATTY ACIDS: SOME ARE NEEDED, SOME ARE NOT

When we think about fat, we typically think of *fatty acids*. You might have heard of saturated and unsaturated fat. Those are fatty acids and they are the smallest fats you can find.

From a chemical point of view, fatty acids are molecules with a tail. These tails (or chains) are covered in hydrogens. The extent to which hydrogens cover the fat's tail is called *saturation*. Saturated fats are completely covered by hydrogens, while unsaturated fats are only partially covered with hydrogens. Why is this important? Because depending on their level of saturation (e.g., how many hydrogens they carry), these fats serve very different roles in your brain—and have profoundly different effects. In other words, a pound of bacon is not nearly the same as a pound of tofu or a pound of olive oil.

Let's begin with saturated fats. We are all familiar with the look and feel of saturated fat—whether we know it by the layer of cream that forms atop full-fat farm milk, the white ribbon that edges a slice of prosciutto, or the marbling on our favorite cut of beef. These fats are quite rigid, which is why butter is solid at room temperature.

On the other end of the spectrum are unsaturated fats. These fats are quite supple but also very delicate and easily damaged. For example, many vegetable oils rich in unsaturated fats, like olive oil, become rancid very quickly if exposed to light or heat, or when they're not stored properly. Another important thing to know is that there are different types of unsaturated fats: *mono*-unsaturated (with one free hydrogen "spot" in its fat chain) and *poly*-unsaturated (with many free "spots").

Monounsaturated fats are abundant in oils such as olive oil, in several nuts and seeds, and in high-fat fruits like avocados, as well as whole milk, wheat, and oats.

Polyunsaturated fats (or PUFAs) are found mostly in oils from plant and marine sources, especially in fatty fish like salmon, algae, and some nuts and seeds. You have probably heard of PUFAs and have seen them encapsulated in the light yellowish liquid of fish oil supplements, labeled as omega-3 fats, or advertised in your breakfast cereal fortified with omega-3s.

Saturated and unsaturated fats are an excellent example of how, even though the brain contains both, it doesn't actually need all of them replenished.

Remember how selective the brain is relative to import/export? Saturated fat is not on the brain's grocery list. Contrary to what many diet books preach, the brain is able to make as much saturated fat as it needs locally, and as such, doesn't require restocking. When you eat a meal that is full of saturated fat (barbecue spareribs or a piece of cheese), the brain might take up a little bit, but for the most part, uptake is minimal after adolescence. The brain needs saturated fat only while it's growing new brain cells, which corresponds to the period of time between infancy and adolescence. After our teenage years we pretty much have all the brain cells we will ever have, and so our brains no longer need saturated fat. It is at that point that the brain's "saturated fat gates" shut down. Afterward, with a few exceptions, saturated fat can't even get inside the brain.

Now for the exceptions. There are a few very specific types of saturated fats that can still enter the adult brain, when needed. Their tail has to be fairly short, like the *butyric acid* found in whole milk, or of medium length, like the *myristic acid* found in coconut oil. Otherwise, the vast majority of saturated fat found in the brain is actually produced inside the brain alone and is not absorbed from our diet. It *looks* like the same fat that we find in meat and cheese, but it doesn't come from these foods. The same applies to monounsaturated fat, which is largely "homemade" on the brain's premises.

Since the brain's got these fats covered, they are referred to as *non-essential fats*. It's as if, in a dialogue that's been going on for millions of years, the brain is telling the body: "Don't worry, I got this. Please bring me the *other* fats I need instead."

So what are these other fats that the brain needs? PUFAs, of course. The rarest and most precious of brain-essential fats.

PUFAs are the only kinds of fat the brain cannot make on its own and, at the same time, determinedly craves. This is especially true for the omega-3s and omega-6s found in fish, eggs, nuts, and seeds. Does this remind you of our early ancestors' diets before meat arrived on the menu? Indeed it does.

PUFAs are the most abundant fatty acids in cell membranes throughout the brain. The brain is specifically designed to collect these fats through dedicated gateways located within the blood-brain barrier. Therefore, a large number of PUFAs are constantly flowing inside the brain—or at least they would if we ate the right foods. These fatty acids are in such immediate demand that as soon as they arrive at the brain, they are consumed accordingly. In fact, the brain needs them to form larger and more complex fats—the phospholipids and sphingolipids mentioned earlier.

Since PUFAs are so crucial to brain health, let's look into these brain-essential fats in more detail.

THE GOOD: THE OMEGA FAMILY

Among all possible types of PUFAs, two varieties, omega-3s and omega-6s, are the best known for promoting brain health. It is definitely worth a concerted effort to eat both of these PUFAs in our daily diets, as they serve very different functions.

Omega-6s are generally regarded as having *pro-inflammatory* properties. This is a big deal. In biology, inflammation is not a swollen thumb. Inflammation means how activated your immune system is, and therefore how well protected you are from danger. Omega-6s participate in this process by encouraging our bodies and brains alike to mount an inflammatory response in the case of a wound or an infection. The omega-3s are needed to turn down this response once the danger is no longer present. As such, they are regarded as being *anti-inflammatory*.

Countless scientific studies have demonstrated that the balance of these two PUFAs is essential for proper neuron communication as well as a means to maintain a healthy immune system. A disruption to this balance of pro- and anti-inflammatory agents could lead to either unnecessarily sustained inflammation, which can damage the brain in the long run, or a reduced capacity to fight off disease, an obvious liability.

This balance directly depends on our food choices.

Research has determined that a ratio of two-to-one (twice the amount of omega-6 to omega-3) is an ideal balance to shoot for. However, there are some estimates that Americans consume *twenty or thirty times* more omega-6s than omega-3s, making the typical American diet highly inflammatory in nature.

Worse still, current dietary recommendations for omega-3 and omega-6 PUFAs strongly favor the latter. The U.S. Food and Nutrition Board has established dietary reference intakes (DRIs) for many nutrients identified as essential to health, including PUFAs. DRIs tell you how much of a nutrient you should consume every day to avoid both deficiencies and excesses that can lead to toxicity. At present, the recommended dose for an adult man is 1.6 grams of omega-3s and between 14 and 17 grams of omega-6s per day, and that for an adult woman is 1.1 grams of omega-3s and 11 to 12 grams of omega-6s. This is ten times the amount of omega-6 to omega-3!

By eating too many omega-6-rich foods and too little omega-3s, we are putting ourselves at risk for many diseases that are associated with excess inflammation in our bodies. These diseases include atherosclerosis, arthritis and vascular disease, autoimmune processes and tumor proliferation, and last but not least, neurological disorders such as Alzheimer's. We need to take this matter into our own hands and make sure our diets are supportive of the important and delicate two-to-one balance.

What are our options?

Starting with the omega-6s, foods that are naturally rich in these fats, and therefore need to be drastically reduced, are fatty animal foods such as bacon and chicken fat as well as vegetable oils from grape seeds, canola, corn, peanuts, and sunflower seeds (Table 2).

As for the omega-3s, there are three major types that come from different foods. These are *alpha-linoleic acid* (ALA), *eicosapentaenoic acid* (EPA), and *docosahexaenoic acid* (DHA). ALA comes from plant sources, especially flaxseeds, walnuts, chia seeds, and wheat germ, along with some sea vegetables such as spirulina. DHA and EPA are plentiful in fish oils instead of vegetable oils. The richest food sources are cold-water fish like salmon, mackerel, and cod. But the real star is caviar. Black caviar has a higher density of brain-building DHA than any other food. One ounce of caviar contains three times the amount of DHA found in the highest-quality salmon. Additionally, caviar is an excellent source of memory-boosting choline, which makes for a perfect brain combo. These are the foods we need to provide our brains with on a daily basis.

Omega-3 PUFA (g/100 g of product)			
PLANT SOURCE	ALA	ANIMAL SOURCE	DHA + EPA
Flaxseed oil	52.8	Caviar, black	6.8
Flaxseeds	22.8	Salmon roe	6.7
Hemp seeds	12.9	Wild salmon	2.2
Butternuts, dried	8.7	Herring	2.0

PLANT SOURCE	ALA	ANIMAL SOURCE	DHA + EPA
Chia seeds	3.9	Mackerel	1.9
Black walnuts	3.3	Sardine	1.7
Soybeans (raw)	3.2	Anchovy	1.5
Oats, germ	1.4	Sardine, canned	1.0
Spirulina	0.8	Trout	0.9
Wheat, germ	0.7	Shark	0.8

Omega-6 PUFA (g/100 g of product)			
PLANT SOURCE	OMEGA-6	ANIMAL SOURCE	OMEGA-6
Grapeseed oil	70	Turkey fat	21
Sunflower oil	66	Chicken fat	19
Wheat germ oil	55	Duck fat	12
Corn oil	53	Lard	10
Soybean oil	51	Pork belly	5.2
Sesame oil	41	Bacon	4.5
Walnuts	38	Egg yolk	3.5
Mayonnaise	37	Chicken	3.1
Peanut oil	32	Frankfurter (beef and pork)	2.3

Table 2. Top ten food sources of omega-3 and omega-6 PUFAs, ranked by PUFAs' density, which is a convenient way to compare the relative amounts of PUFAs available to the brain in each food.

A large body of epidemiological studies identified omega-3s as the number one nutrient to fight age-related cognitive decline and dementia. For example, a landmark study of six thousand participants ages sixty-five years or older showed that people who consumed low quantities of omega-3s had a 70 percent higher risk of developing Alzheimer's than those who consumed more omega-3s. Specifically, people who consumed less than 1 gram of omega-3s per day had the highest risk, while those who ate more than 2 grams were unlikely to develop dementia at all. Moreover, even among people who did not develop dementia, a lower intake of omega-3s directly impacted their ability to remember details, switch focus, and manage time.

As several other studies replicated these findings, a consensus was reached: people who consume omega-3-rich foods on a regular basis remain mentally clearer and have a lower likelihood of developing cognitive deterioration than those who consume less of these healthy fats.

Perhaps even more fascinating is that the beneficial effects of omega-3 consumption are evident in brain scans as well. As we age, our brains naturally lose some volume. But studies of thousands of dementia-free elderly showed that those who didn't consume enough omega-3s in their diets had *accelerated* brain shrinkage, as shown on MRIs. Their brains, and especially their hippocampus (the memory center of the brain), lost neurons at a rate equivalent to two extra years of aging. In other words, a diet poor in omega-3s, especially DHA, increases the speed at which the brain ages! Some specifics: people who ate less than 4 grams of DHA a day showed the highest rates of brain shrinkage, whereas those who consumed 6 grams or more had the youngest-appearing brains.

While all of this points to the protective effect of omega-3s on brain aging, randomized clinical trials have still failed to show significant changes in cognitive health as a result of fish oil supplementation. Why is this?

Rather than being disappointed by the negative results, there's actually a lesson to be learned. First, clinical trials were based on supplements. But research shows that the protective effects of omega-3s are particularly strong in people who derived their fats from natural sources like fish *rather than from supplements*. Further, most clinical trials to date used fairly low doses of omega-3s, generally about 2 grams a day—less than *half* the minimum dose found to slow down brain shrinkage in the MRI studies mentioned earlier. This might be related to current dietary guidelines recommending even lower doses of omega-3s, especially for women.

In the end, the research studies offer a much better, brain-specific rule of thumb. The goal is to eat *at least* 4 grams of our brains' coveted fat every day to keep our brains young and working to the fullest.

It really doesn't take much to reach our brain-fat goals. Just 3 ounces of wild Alaskan salmon (that's a small piece) provides almost 2 grams of omega-3s, most of which is DHA. Pair that with a handful of almonds (or perhaps a scoop of caviar) and you're good to go.

As for omega-6s, you need very little: a few drops of grapeseed oil or a handful of peanuts is all your brain needs for the day.

Easier said than done. Unfortunately, the typical Western diet tends to include exactly the opposite proportions of these foods. There are several ways to reverse this balance to increase our consumption of omega-3s. A simple way to do that is to swap out some omega-6-rich foods, such as the peanuts in our peanut butter, for other nuts higher in omega-3s, like almonds and walnuts.

Another good idea would be to increase our consumption of cold-water fish by eating it in lieu of pork, beef, and other fatty meats. If we compare a portion of fish and beef possessing the same amount of protein, the fish is rich both in protein *and* omega-3s, while the meat provides more omega-6s than omega-3s.

When properly prepared, fish is a delicious and healthy alternative our brains need more than anything else. The Mediterranean diet, common in my home country of Italy and throughout much of Europe, features fish much more than is traditional here in the States. What most of us really appreciate is that fish tastes like the sea—with its briny flavor, the slight iodine tang to it, so reminiscent of the jeweled turquoise waters of the Mediterranean Sea. Red snapper, mussels, *vongole* (clams), *orata* (sea bream) . . . they all taste like a beach holiday to me.

Every sea has its own forte. A grilled fresh-caught *branzino* (sea bass) is a Mediterranean delicacy, as are oysters in France and wild salmon and herring in northern Europe. Japanese cuisine, particularly sushi, has now become popular throughout the world, and eating raw fish has replaced the traditional slice of pizza as the dinner of choice in many households, bringing an unprecedented amount of healthy omega-3s to American tables.

For you traditionalists who really love your steak, invest in grass-fed beef, which is far richer in omega-3s and void of the GMO and hormonal feeds present in lower-quality beef.

Another perhaps unexpected way to replenish our reserves of brain-essential fat is to eat foods rich in phospholipids. Even though you may not have heard about them until now, phospholipids are present in your entire body and much more so inside your brain. This particular type of fat is crucial to providing all brain cell membranes with shape, strength, and elasticity. As such, it is needed for all mental operations and thoughts to take place. Phospholipids are made up mostly of omega-3s. As a result, foods that contain phospholipids are at the same time excellent sources of omega-3s. What happens is that your body breaks down these bigger fats into smaller units, releasing their omega-3s first in the circulation and then inside the brain. Phospholipids-rich foods include all fish, crustaceans like crab and krill, and eggs. The egg yolk in particular packs the biggest punch, with over 10,000 mg of phospholipids per 100 grams of product—and also 230 mg of omega-3s. Think about it next time you're about to order an egg white–only omelette—you're throwing the baby out with the bathwater. Of all animal foods available to us, eggs are hard to beat for brain nutrition. In a way, eggs are to animals what berries are to plants—over the course of thousands, if not millions, of years, eggs have been optimized to produce and sustain life by providing the offspring with nutrition, structure, and protection, all in each egg. That's because eggs contain all sorts of nutrients that are needed not only for bones and muscles to grow but also for the neural tube to develop into the brain and spinal cord. Memory-forming choline, omega-3 fatty acids, complete protein, a wide range of vitamins and minerals, and even disease-fighting antioxidants like lutein and zeaxanthin—the egg has it all because the developing organism needs it all.

When we talk about eggs, we typically refer to chicken eggs. But the same principle applies to all egg-laying birds, from quail to ostrich. I recommend that you explore all available options, as diversity is key in neuro-nutrition. And let's not forget about fish. Fish eggs, aka caviar or roe, are the best treat for our big, hungry brains. Even Picasso expressed his gratitude to benefactors by presenting them with caviar, and did them a greater favor than he knew. Caviar is the most potent phospholipid source available, not to mention that one tablespoon alone contains more than 1 gram of high-quality omega-3s.

Additionally, the vegetable kingdom is also rich in phospholipids. Sweet peas, cucumbers, and tapioca (a staple of South American cuisine) are unexpected yet plentiful sources. Grains like oats, whole wheat, and barley, as well as soybeans and sunflower seeds, are also good sources. By eating these foods, you'll help your brain obtain all the "brain fat" it needs. This is especially helpful if you have medical reasons or allergies requiring that you steer clear of fish, or if you are vegan/vegetarian. There are also other ways to increase your omega-3s. For example, you can eat more foods that contain ALA. ALA is found in considerable amounts in plant sources, especially oils from flax, chia, hemp, and sunflower seeds (Table 2), making them an attractive option for vegetarians. The catch is that our brains must convert ALA to DHA to meet its needs, and 75 percent of ALA is lost in the process.

High-quality fish oils containing DHA are a suitable alternative, easily available in the form of supplements or as ingredients in a wide variety of fortified foods, from DHA-fortified milk to enriched eggs and bread. Many people who take fish oil supplements remark on the pill's lingering, fishy aftertaste. If that bothers you, try a vegan source of omega-3s derived from marine plants, such as high-purity algae, instead. Another advantage of these vegan supplements is that they are free from the contaminants and environmental pollutants that might be present in some fish oils.

Finally, although results are not always consistent, some studies have shown that a higher consumption of monounsaturated fats such as those found in olive oil and avocados correlates with better cognitive functioning and a lower risk of dementia in the elderly. In those studies, people who consumed at least 24 grams of these fats a day had an 80 percent reduced risk of Alzheimer's as compared to those who consumed 15 grams or less. Monounsaturated fats are renowned for their beneficial effects on heart health, and in this case, what's good for the heart is good for the brain. The good news is that it takes only 2 to 3 tablespoons of olive oil to reach the desired brain-protective daily dose.

To sum it all up, omega-3s, omega-6s, phospholipids, and, to some extent, monounsaturated fats are the good guys the human brain has chosen as its allies—not to mention suppliers—over the course of millions of years. At the same time, your brain has also made a few enemies.

THE BAD: SATURATED FAT

Even though this seems to be relatively unclear to the public, saturated fat tops the list as public enemy number one. While some doctors support and even encourage unlimited consumption of foods rich in saturated fat, it is well established in the scientific community that high levels of these fats have a negative impact on our mental capacities, increasing our risk of dementia.

As described earlier, the brain doesn't consume much saturated fat after adolescence. However, a high intake of saturated fat from the diet can cause inflammation throughout your entire body and reduce oxygen flow specifically to the brain. The brain is a glutton for oxygen, so it's possible that even a slight lack of circulation could affect its performance. Additionally, too much saturated fat can increase the risk of heart disease and type 2 diabetes—which in turn raises one's risk of dementia.

One example among many. A study of over eight hundred elderly participants showed that those who consistently ate the most saturated fat had up to four times the risk of developing cognitive deterioration as they aged as compared to those who ate the lowest amount. Specifically, people who consumed more than 25 grams of saturated fat per day were much more likely to develop dementia in the years to come than those who ate half that amount (13 grams per day).

In practical terms, six slices of bacon contain 25 grams of saturated fat. So cutting down your bacon from six slices to three can cut your risk of dementia in four.

Subsequent studies have shown that even though 13 grams a day are less bad for you than 25, they are too many anyway. For instance, a study of six thousand elderly showed that those who ate 13 grams of saturated fat (or more) daily were almost twice as likely to develop cognitive impairment as those who ate less than 7 grams per day. Now we're down to one and a half slices of bacon.

A number of well-controlled studies have confirmed these findings: high saturated fat increases the rate of decline in cognitive abilities as we age. So while the body certainly needs some saturated fat to stay healthy, scientists agree that as far as the brain is concerned, the less saturated fat, the better.

Currently, most nutritionists recommend aiming for a diet that includes no more than 5 percent to 6 percent of calories from saturated fat a day. For example, if you are on a 2,000-calorie-a-day diet, no more than 120 of these calories should come from saturated fats. That's about 13 grams a day, or three slices of bacon, which is consistent with the studies mentioned earlier. As you can imagine, most people eat much more saturated fat than a few slices of bacon a day. So if you want to maximize your brain's performance, while at the same time minimizing your risk of dementia and heart disease, I would recommend eating less than 13 grams a day, or even better, half of that—while also focusing on healthier, high-quality sources.

If you're wondering which meat and dairy products are safe, here's my rule of thumb. Fresh, organic fatty foods are much better for you than processed foods of any kind. When you're eating animal foods, make sure you eat organic free-range eggs, chicken, and turkey and lean grass-fed cuts of beef *instead* of commercially raised meat, pork, and bacon. Ditto for all processed meats, whether processed cold cuts of ham or mixed meats like pastrami. These foods are not good for you, especially for your brain. We'll talk a whole lot more about this in Step 2: Eating for Cognitive Power.

With respect to dairy products, fermented, organic whole milk products like yogurt and kefir are definitely our best bet. Besides providing a better ratio of unsaturated to saturated fats, they have the added benefit of containing live active bacteria that support the digestive and immune systems alike. Instead, processed foods like American cheese, string cheese, sweetened yogurt, commercial ice cream, pudding, and most milk beverages are downright bad for you and should be restricted, if not completely eliminated from your diet. Not only are these foods high in the kind of saturated fat the brain has no use for, but they are also high in multiple harmful ingredients, not the least of which are trans fats.

THE UGLY: TRANS FAT

We all have heard about trans-saturated fats (aka trans fats) and how terrible they are for the body. Most doctors consider trans fats the worst type of fat you can possibly ingest. This kind of fat has been making headlines of late for its dangerous effects on health, and it is increasingly recognized as a brain enemy as well.

A large body of literature indicates an association between consumption of trans fats and increased risk of cognitive decline and dementia in old age. In fact, it takes very little trans fat in the diet to develop cognitive impairments. Across several studies, people who consumed 2 or more grams of trans fats a day had twice the risk of those who ate less than 2 grams. Fairly disheartening is the realization that most people in those studies ate *at least* 2 grams a day, with the majority of participants eating more than double that on a regular basis.

But what exactly are trans fats, and where are they hiding?

Trans fats are created via an industrial process called *hydrogenation*, by which hydrogen is added to otherwise healthy unsaturated vegetable oils, thereby chemically "saturating" them. Manufacturers do this to create a specific consistency in products, one that is nearly solid at room temperature but melts upon baking or heating. For example, canola and safflower oils are artificially hydrogenated to create margarines and soft spreads. These so-called partially hydrogenated oils are less likely to spoil and less prone to rancidity, granting foods a longer shelf life. Some restaurants use these oils in their deep fryers because they don't have to be changed as often as do other, healthier oils. Clearly, trans fats are convenient, not to mention cheap.

Unfortunately, trans fats have many adverse health effects, from raising cholesterol and triglyceride levels in the bloodstream, to globally promoting inflammation throughout our bodies. This increases our risk of cardiovascular disease and stroke, and in turn, our risk of dementia. As of today, partially hydrogenated oils are no longer "Generally Recognized as Safe" (GRAS) in human food. Several countries, like Denmark, Switzerland, and Canada, have reduced and even restricted the use of these fats in food service establishments. There is hope that continued research will soon lead to a final and complete ban of all trans fats from the American diet. In the meantime, we need to protect ourselves and safeguard our brains from any food products that include trans fats.

They are relatively easy to spot. First of all, they are almost always found in processed foods. The definition of what constitutes a processed food can vary, but generally speaking, it's a food packaged in boxes, cans, or bags. Also, these foods have a very long shelf life. Some canned soups can last up to four years. That can't be natural, can it?

It appears easier than ever to determine the amount of trans fats in any packaged food by checking its nutritional label. A typical Nutrition Facts panel will list not only the serving size and calories per container but right underneath that you will find Total Fats. The Total Fats listing is then broken down into saturated fats, cholesterol, and trans fats. You want anything you put in your cart to read trans fats: zero/"0" grams.

But there's a catch. Due to some latitude in current regulations, even foods appearing to be trans fat–free might instead contain up to 0.5 grams of trans fats per serving. In other words, if a food contains 0.49 grams of trans fat per serving, the company is allowed to list it at "0" grams.

But how many people only eat just one serving of anything? Let's say you eat two servings of such a food (such as a "butter spread," where two servings could equal as little as 2 teaspoons). Doing so, you'll actually be eating almost 1 full gram of trans fat—all the while thinking you didn't have any at all. One example among many is the product Land O'Lakes Fresh Buttery Taste Spread. The Nutrition Facts panel proudly states "0" trans fats. But the ingredients list tells a whole different story. Several trans fats feature in the smoothness of the spread, such as partially hydrogenated soybean oil, hydrogenated soybean oil, and hydrogenated cottonseed oil. If you can't believe it, google it.

The bottom line is this: the more packaged and processed foods you consume on a regular basis, the more hidden trans fats you are probably consuming, and the higher your risk of getting sick. My recommendation is to take a careful look at the ingredients list on the food packages you pick up. Watch out for any of the following: hydrogenated fats (like the ones mentioned above), partially hydrogenated fats and oils (also known as PHO), and vegetable shortening (or just shortening). The most common PHO include partially hydrogenated soybean oil, cottonseed oil, palm kernel oil, and vegetable oil. If the list includes any of these substances, you are better off putting the package down.

Typical processed foods that are high in trans fats are baked goods like commercial doughnuts, cakes, pie crusts, biscuits, and frozen pizza, as well as many snack foods such as cookies and crackers. Then there are all the margarines (stick or spread), along with many other spreadable or "creamy" products, which are by definition made of hydrogenated or partially hydrogenated oils. Trans fats are even added to most coffee creamers. A special note to moms: you might be inadvertently serving your children a considerable amount of toxicity atop their birthday cakes, since ready-to-use frostings are plied with trans fats. So take a close look at your shopping cart. Many of the foods you stock in your pantry are probably laden with a substance that we must now begin to avoid at all costs.

CHOLESTEROL: FRIEND OR FOE?

Cholesterol has become such a hot topic that we even find ourselves talking about our cholesterol levels at parties. Is cholesterol a good thing or a bad thing?

What might come as a surprise is that brain cholesterol is very different from the cholesterol we generally talk about. When your doctor tells you that your cholesterol is too high, they are referring to the cholesterol in your blood. Your levels of blood cholesterol are at least in part determined by how many cholesterol-rich foods you eat, like meat, eggs, and some dairy products.

However, that cholesterol has *nothing* to do with the cholesterol inside your brain.

Unlike any other organ in the body, the brain makes its own cholesterol. All of it. Perhaps even more surprisingly the majority of brain cholesterol is produced during our very first weeks of life, when our neurons are growing at light speed and need the extra structural support. The brain continues to make cholesterol at a lower rate throughout adolescence, but once that process is complete, it slows down to make very little during adulthood.

By the time we are past our teenage years, we have all the brain cholesterol we will ever have. In order to preserve its own cholesterol, the brain completely seals it away from the rest of the body. As an additional safety measure, dietary cholesterol (from the foods we eat) is not on the list of nutrients that are allowed to cross the blood-brain barrier. There are simply no passageways to let it in.

Consequently, there is no connection between the amount of cholesterol you eat and your brain's function. No matter how many eggs you put in your omelette or how much bacon you've had on the side, the cholesterol from those foods will not benefit your mental capacities—although it might eventually clog your arteries.

That way, cholesterol from the diet can affect the brain in as much as it affects heart health. When our hearts are not running at their optimal levels, our brains suffer as well. Here's the story.

The cholesterol in your body needs help to move about. Help is provided in the form of *lipoproteins*, the private drivers of your body's fat-transport system. You might recognize them from your most recent blood work. They are split into two groups: low-density lipoproteins (LDLs) and high-density lipoproteins (HDLs). When doctors talk about "bad cholesterol" and "good cholesterol," they are referring to these molecules.

LDLs carry cholesterol to specific organs. Sometimes this process doesn't run as smoothly as it should. Because of genetic inheritance or other medical conditions, LDLs can end up dropping their passenger at the wrong destination—inside your artery walls, for instance. As this process continues, a buildup of cholesterol can occur, beginning to cause a thickening along the artery. This thickening is called a plaque, or *atherosclerotic plaque formation*. As these plaques increase, they can eventually clog the arteries, causing heart attacks, strokes, and other serious medical problems. Instead, HDLs pick up cholesterol and shuttle it to the liver, where it is either eliminated or converted into hormones for other uses. This is why, even though LDL and HDL are not actually cholesterol, they got labeled "bad LDL cholesterol," becoming the villain we need to get rid of, and "good HDL cholesterol," the superhero we seek.

Regardless of medical terminology, in order to protect your heart and brain alike, it is crucial to keep your total cholesterol low, particularly in conjunction with high HDL and low LDL.

A large number of studies agree that if you have high blood cholesterol (*hypercholesterolemia*) in midlife, you are at increased risk for developing dementia later in life. For reference purposes, the term "high cholesterol" indicates a blood cholesterol level higher than 240 mg/dL. Several studies with up to ten thousand elderly participants found that those with high cholesterol in midlife had almost three times the risk of cognitive issues and dementia later in life as compared with those who had normal cholesterol levels. But when researchers looked at the risk of Alzheimer's in particular, the healthy limit was even lower. A cholesterol level of 220 mg/dL, which is considered borderline high by current standards, was high enough to nearly double the risk.

What are we to do to keep our cholesterol in check?

Traditionally, if you had high cholesterol, your doctor would advise you to reduce consumption of cholesterol-rich foods such as eggs and cheese. However, it is becoming increasingly accepted that cholesterol from our food doesn't increase blood cholesterol levels as much as once thought. The truth is, up to 75 percent of cholesterol in your body is produced by the body itself, and only 25 percent or so is derived from the diet. This is because the body tightly regulates the amount of cholesterol in your blood via its strict internal controls to make sure that you do not absorb too much cholesterol from the foods you eat. In other words, eating cholesterol isn't necessarily going to give you a heart attack.

However, other nutrients can push your cholesterol levels up to the sky. As puzzling as this might sound, it is consuming saturated and trans fats that seems to affect cholesterol levels more than consuming cholesterol itself. So if you need to lower your cholesterol, you are better off reducing consumption of those *other* fats we talked about earlier. We have already clarified that trans fats should be eliminated altogether from your kitchen. Foods high in saturated fat should also be limited. The tricky part is that the majority of cholesterol comes from foods that also contain a good amount of saturated fat, like fatty meats, pork, poultry, and dairies, which makes it hard to eat one but not the other.

Fish, shellfish, and eggs are exceptions to this rule, as they are high in cholesterol but lower in saturated fat, and therefore not as harmful as previously thought. For instance, clinical trials reported no association between eating eggs and the risk of heart disease. One more good reason to ditch the egg-white omelettes and start eating yolks again. However, if this news inspires you to pay a visit to your local diner for a ten-egg scramble, you are taking this encouragement too far. Healthy egg consumption means no more than a couple of eggs a week. Again, mind the old adage: "Everything in moderation."

Also keep in mind that not everyone responds to dietary cholesterol or saturated fat in the same manner. If we took a group of people, fed them a diet high in those fats, and measured how much cholesterol they produce, we'd see a wide range of responses. Some will make more, some will make less. If you have high cholesterol, or a family history of heart disease, you might want to talk to your doctor about testing your own individual response to eating fatty foods. If consuming saturated fat does increase your total or LDL cholesterol, you have plenty of good reasons to change your diet.

BRAIN AND BODY—ARE THEY ON THE SAME PAGE?

Now that we've reviewed the major types of fats and whether or not they belong in a brain-healthy diet, let's sum it all up. Certain dietary fats such as PUFAs are an excellent example of how what's good for the brain is good for the body. The brain needs PUFAs (especially PUFAs from fish and eggs) for proper structure and function. The rest of the body needs PUFAs as well, and can even burn these fats to make energy. For instance, fatty acids like PUFAs are the heart's main source of fuel. So you want to make sure you eat enough of the PUFA-rich foods listed in this chapter to keep both your brain and your heart happy.

However, it doesn't necessarily work in the reverse, and not all fats that are good for the body are good for the brain. There are some who recommend eating foods rich in saturated fat and cholesterol to keep your brain healthy. I disagree. In the case of saturated fat, the body alone has the ability to burn this fat to make energy. Likewise, the body alone can use dietary cholesterol to carry out a number of functions—everything from supporting organ membranes to producing several hormones. The brain, on the other hand, has no particular need or use for these fats after adolescence.

That said, the subject of saturated fat is a complicated one and it is important to keep in mind that some saturated fats are better for you than others. As mentioned earlier, occasionally the brain might do well to take up some saturated fats like the ones found in whole milk and coconut oil. Even though your brain doesn't require these fats on a regular basis, it won't hurt to consume them sparingly or on occasion—especially in preference to other, more harmful sources of fat like trans fat. We'll talk more about which fatty foods are best to support both your brain and the rest of your body in the next chapters.

Meanwhile, a note of caution for all new parents out there. Don't assume that what's good for you is good for your two-year-old. A developing brain (which means anywhere between birth to adolescence) needs more saturated fat than an adult brain. In Italy, pediatricians recommend that children eat at least two servings of high-quality meat a week, along with plenty of (organic, fresh) milk and dairy and, of course, fish. I follow this rule with my two-year-old, who adores wild salmon and Parmesan cheese, and also enjoys coconut oil as much as organic grass-fed sweet cream butter. As we'll discuss later in the book, and especially if you have children, of all the criteria to hold for food, the *quality* of the ingredients might well be the most important part.

5

The Benefits of Protein

THE BUILDING BLOCKS OF PROTEIN

Proteins come third in the lineup of top brain-healthy nutrients. Proteins are complex molecules that do most of the work in our cells and are required for the structure, function, and regulation of the brain's networks. They are made up of smaller units called *amino acids* that are attached to one another in smaller or longer chains. The number and sequence of amino acids used to build a protein determine that protein's unique shape and properties.

Amino acids are essential for just about every function that takes place within the body and brain. This includes maintaining healthy tissues, assembling hormones, and powering all sorts of chemical reactions, for starters. But even more important for the brain, amino acids have long resided in the minds of all creatures on Earth. In fact, many of these nutrients act as *neurotransmitters*, the chemical messengers that our brains use for signaling, communicating, and processing information. Neurotransmitters are responsible for how you think, talk, dream, and remember. They spur the impulses that wake you up, make you sleepy, keep you focused, and even cause you to change your mind.

Adequate amounts of amino acids, and therefore of the proteins that supply them, must be provided to the brain on a daily basis for all these cognitive functions to be carried out. However, the human brain has figured out how to manufacture some amino acids on its own, while importing the rest from the diet. So even though all amino acids are needed for *overall* health, the brain alone doesn't require all of them.

Just like fats, amino acids can be *essential* or *non-essential.* Amino acids that can be produced directly by the brain are labeled as non-essential, while those that cannot be made internally and must be supplied by the diet are labeled as essential.

When you eat a meal that contains protein, essential amino acids like *tryptophan* (the molecule found in turkey that facilitates sleep) enter the brain very rapidly thanks to special passageways embedded in the blood-brain barrier. Instead, non-essential amino acids (such as the *asparagine* found in asparagus) are markedly restricted from entry into the brain.

In general, proteins are not hard to come by. Several foods contain complete, well-balanced proteins, which means that they provide all essential amino acids. These foods are generally of animal origin and include fish, milk, eggs, chicken, pork, and beef. Many plant-based foods such as legumes, grains, soybeans, and some nuts and seeds also contain good amounts of protein. In the next pages we'll see which proteins are specifically needed for a healthy, active brain and which foods provide the best amino acid content to support mental clarity and sharpness throughout a lifetime.

STARS, TREES, SPIDERS, KITES, AND CARROTS . . . OH MY!

The most immediate way to appreciate just how intrinsically dependent our cognitive abilities are on the right combinations of brain-essential amino acids is to look at those very mechanisms that enable our thoughts to take shape in the first place. We will now look at how brain cells communicate with one another, transmitting information throughout the brain to form ideas, memories, and feelings—and how much, if not most, of this process is literally based on protein.

If you were to gaze into the brain through a microscope, you would witness a peculiar sight. The human brain is packed with an incredible array of brain cells, each one a different shape and size.

Some might look like stars and others more like trees, each one sporting its own signature branch structure. Another grouping might look more like spiders, while others appear more like kites trailing their long tails. Some even look like carrots, with their leafy tops. These stars, trees, spiders, kites, and carrots are all working diligently to carry messages back and forth, keeping the communication going among all the various brain regions before further branching out to share the information with the rest of your body.

The orchestra of our central nervous system, for which the brain is the conductor, is composed of over 80 billion brain cells called *neurons*. Neurons are unique among all the cells that make up our bodies because of their ability to send signals to other cells no matter how far the distance. It's precisely these varying shapes and sizes that enable them to do so.

Throughout this amazing electrical relay, our neurons do not actually touch. Their bodies, or axons, end in several *synapses*, which are the tiny gaps that separate one neuron from the next and act as connection points. Every neuron has between 1,000 and 10,000 synapses, totaling an astounding 100 trillion connections overall.

Each electrical impulse traverses many synapses as it makes its way through the nervous system. This "neurotransmission" is carried out by neurotransmitters, the chemical messangers of the brain. Neurotransmitters are in charge of dispatching information from cell to cell. The continued, incessant influx of neurotransmitters is precisely what gives birth to our thoughts, memories, and words—and even gives us a good night's sleep—setting the stage for healthy mental functioning. The human brain relies on the well-orchestrated action of over one hundred neurotransmitters, each with its own specialty and particular chemical makeup. Some of these neurotransmitters have a profound effect on our cognitive performance and mental capacities. One, called *serotonin*, influences your emotional stability and sleep patterns, but also memory and appetite. Another, called *dopamine*, is responsible for reward-motivated behavior, movement control, and cravings. Additionally, there's *glutamate*, a neurotransmitter responsible for provoking you to spring into action, and its alter ego, *gamma-aminobutyric acid* (GABA), that inhibits you from acting at all. We can all imagine just how deeply these molecules affect our brainpower and mental abilities.

As it turns out, many cognitive issues arise as a consequence of neurotransmitter abnormalities. For example, a marked reduction of the neurotransmitter serotonin (the one essential for mood, memory, and appetite) is a typical finding in patients with depression, which in turn can affect memory and attention.

What could be causing this depletion of our neurotransmitters?

You guessed it—*a poor diet.*

Our neurotransmitters are created from one source: our food.

A closer look at the neurotransmission process reveals a surprising fact—neurotransmitters are not sitting around waiting to shuttle their next signal. They are actually produced each and every time there is a need to carry one of the brain's various messages. They report for duty when a message presents itself and disappear again once their mission is complete. This incredibly sophisticated and yet delicate process is intrinsically dependent on several nutrients extracted from the foods we eat day after day. Consequently, the production of neurotransmitters inside the brain responds quite rapidly to changes in our diet—especially our protein intake.

SLEEP, DAYDREAM, AND REMEMBER WITH SEROTONIN

Serotonin has been mainly associated with our moods, as this neurotransmitter signals our brains when we are feeling relaxed and happy. When the brain produces low levels of serotonin, happy signals become much less frequent and short-lived, eventually producing diseases such as depression and anxiety. Serotonin is also largely acknowledged for its role in sleep and appetite regulation. Less well-known is that its depletion can also be responsible for some aspects of the memory impairments associated with advancing age and dementia.

Food is the essential impetus for serotonin production. In fact, the production of serotonin in the brain is based on the presence of the amino acid tryptophan, and is critically dependent on how much tryptophan is available to the brain. Tryptophan is an essential amino acid, meaning it cannot be produced in the body at all. Consequently, the only way we can make it available to our brains is through the foods we eat on a daily basis.

What foods provide us with tryptophan? And how can we get enough of it?

According to current dietary guidelines, the average adult, man or woman, needs 5 mg of tryptophan per kilogram of body weight daily. For example, the recommended dose for a 175-pound (79 kg) adult is 395 mg of tryptophan per day.

This brain-essential nutrient is not too difficult to come by. Several foods contain tryptophan, especially those rich in animal- or plant-based protein. But there's a catch: tryptophan generally takes a backseat to other amino acids in terms of how often and how easily it is absorbed by the brain. Additionally, less than 10 percent of all the tryptophan that we import from our diet is used to make serotonin. Consequently, it's even more important that we consume enough tryptophan-rich foods every day to reach a level that makes a difference in our brains.

Table 3 provides examples of common foods that are good sources of tryptophan. It's interesting to note that although popular lore leads us to believe that tryptophan is responsible for the drowsiness we feel after a turkey dinner, turkey ranks quite low on the list of tryptophan-rich foods. So low that it didn't even make it into Table 3.

Chia seeds instead rank at the top. Chia is one of Nature's powerhouse plant-based foods. These tiny brown seeds have long been known for their high nutritional value and ability to provide sustainable energy. *Chia* is the ancient Mayan word for "strength." In fact, ancient Aztec and Mayan warriors literally survived on rations of chia at times, as did long-distance runners and messengers. Two tablespoons of chia seeds contain over 200 mg of the tryptophan your brain needs to make serotonin, as well as a good amount of omega-3 PUFAs, minerals, and fiber.

Additionally, plant-based foods like raw cacao (chocolate!), wheat, oats, spirulina, and sesame and pumpkin seeds figure prominently among the richest natural sources of tryptophan on the planet. Animal foods like milk, yogurt, chicken, and fish like tuna and salmon are also a good start. In particular, yogurt is an excellent protein source with many health benefits, especially for the digestive system. However, I am not talking about the sugary, creamy, fruity, yogurt-like substances you find in bright packages on supermarket shelves. Those products are full of artificial sweeteners and colorants, not to mention preservatives, and will do nothing to help your brain stay healthy over time. When I say yogurt, I mean organic, plain, tart yogurt—preferably full-fat and from goat's milk for an extra protein kick. If you don't like tart flavors,

I recommend you sweeten the yogurt yourself with raw honey, maple syrup, or fresh fruit.

Food item	Unit (imperial)	Unit (metric)	Tryptophan (mg)	Sum of CAAs (mg)	Tryptophan/ CAA ratio
Chia seeds	1 ounce	28 g	202	1,270	0.159
Milk, whole	1 quart	946 mL	732	8,989	0.081
Sesame seeds	1 ounce	28 g	189	2,330	0.081
Yogurt, whole, plain	1 cup	245 g	49	3,822	0.078
Pumpkin seeds	1 ounce	28 g	121	1,615	0.075
Prunes (dried plums)	1	26 g	2	27	0.074
Seaweed, Spirulina	1 ounce	28 g	260	3,768	0.069
Raw cacao	1 ounce	28 g	18	294	0.061
Wheat bread	1 slice	50 g	19	317	0.060
Edamame	1 cup	118 g	236	2,354	0.057

Table 3. Top ten tryptophan-rich foods, ranked by the ratio of tryptophan to competing amino acids (CAAs). This ratio provides the best indication of tryptophan availability for transport across the blood-brain barrier for use in serotonin synthesis.

In general, tryptophan is found in many common foods, so deficiencies are unlikely. However, too little protein in the diet can cause a deficit. If you are vegan or eat very little animal protein, your tryptophan levels might be low. In that case, make sure you are eating enough of the richest vegetarian sources. Additionally, tryptophan (in the form of 5-Hydroxytryptophan, or 5-HTP) can be taken as dietary supplements.

Here's another trick. Research has shown that eating carbohydrates with or immediately after tryptophan-rich foods helps increase its absorption, thereby increasing serotonin production. This observation is at the heart of many nutritionists' recommendations to eat some carbs for dinner to promote tryptophan absorption and facilitate sleep. When I was little, my *mamma* would make me a cup of warm milk with honey before bed to help me fall asleep. Little did she know that this simple practice helps tryptophan get through to the brain, boosting serotonin production and, therefore, sleep.

Finally, sometimes the problem is not tryptophan availability as much as a vitamin deficiency. Although tryptophan is indispensable for serotonin production, vitamin B6 is also needed to convert tryptophan into serotonin. As we'll see in the next chapters, this vitamin truly is a brain must, as it is crucial not only for serotonin production but also for several other neurotransmitters.

DOPAMINE: MIND YOUR STEP

Although serotonin gets a lot of mainstream attention and most of the hype, dopamine has been gaining increasing popularity due to its role in cognitive function. Dopamine impacts our reward, motivation, and attention functions, as well as problem solving and motor control, not to mention allowing us to feel pleasure in the first place.

Dopamine abnormalities are involved in several medical conditions including Parkinson's disease, attention deficit hyperactivity disorder (ADHD), schizophrenia, and drug addiction.

This neurotransmitter is manufactured in the brain by breaking down the amino acid *tyrosine.* Tyrosine is a non-essential amino acid, which means our bodies are capable of producing this nutrient on their own. But there's a catch. Tyrosine needs to be produced from another amino acid called *phenylalanine,* which happens to be an essential amino acid, meaning it must be obtained from our diet instead. What this all boils down to is that without eating foods containing phenylalanine, we might as well kiss good-bye our pleasurable strolls in the park or playing our favorite games—let alone enjoying the thrill of winning them.

In order to be sure you're getting enough of these nutrients, keep an eye on the recommended dose of phenylalanine and tyrosine. We all need 33 mg per kilogram of body weight per day. For example, the recommended dose for a 175-pound (79 kg) adult, man or woman, is 2.6 grams per day.

As shown in Table 4, phenylalanine is found in many high-protein animal products such as chicken, beef, eggs, and fish (particularly fatty fish like salmon, striped bass, and halibut), as well as dairy products such as milk and yogurt. Plant-based foods are also excellent sources, particularly legumes like soybeans and peanuts, nuts like almonds, and once again, chia, pumpkin, and sesame seeds. Oh, and don't forget spinach.

Deficiencies of both amino acids are rare, but any medical condition or dietary regimen that reduces intake of these amino acids would consequently affect dopamine production in the brain.

Animal sources	Phenylalanine (mg/100 g food)	Plant sources	Phenylalanine (mg/100 g food)
Parmesan	1870	Soybeans	2122
Cheddar	1390	Peanuts	1290
Chicken	1310	Chia seeds	1028
Beef steak	1210	Almonds	980
Organ meat	1200	Sesame seeds	959
Pork, leg	1030	Pumpkin seeds	924
Prawn	910	Walnuts	540
Codfish	790	Chickpeas	460
Salmon	775	Lentils	400
Seabass	760	Kidney beans	350

Table 4. Top ten food sources of phenylalanine, ranked by phenylalanine density, which is a convenient way to compare different foods and the relative amounts of phenylalanine they make available to our brains.

Supplements for these amino acids are also available. In this case, I recommend supplementing with the natural form L-phenylalanine, as opposed to D- and DL-phenylalanine, which are synthetic forms. Always ask your health-care provider for advice before taking any supplements.

GLUTAMATE: READY, SET, GO . . . OR WAIT, HIT THE BRAKES!

I once read an interesting analogy about how the brain handles decision making. Picture this: someone walks into the room holding a steaming slice of hot pizza, and you think, "Wow, I want that pizza and I want it NOW!" The neurons responsible for initiating your acting on that thought are about to send impulses skyrocketing across your brain to achieve your aim, when suddenly another thought comes to mind. "Maybe I shouldn't eat pizza because, truth be told, I *am* trying to lose weight." A different group of neurons immediately sends a counter impulse to put everything on hold. But then you smell that pizza again and decide that you'll eat a little bit of it anyway. Just as the Team Pizza neurons start firing in all directions ready to celebrate their win, you discover that you don't have any cash on you. The Team Don't Do It neurons respond by putting an end to the discussion once and for all. You sigh and go about your business, if a little disappointed at the outcome.

The point is that your brain can both initiate an action and constrain the same action, all at an incredible speed. This is possible due to the fact that we possess different neurotransmitters to achieve differing goals. There are *excitatory* neurotransmitters, which push neurons to transmit signals, and *inhibitory* neurotransmitters, which make signal transmission less likely. Consider it the yin and yang of neurotransmission—or a devil and an angel on each of your shoulders.

Glutamate is the main excitatory neurotransmitter of our entire nervous system from head to toe. It's the one that makes you reach for your wallet to buy the pizza. To get a better sense of how important this little molecule is, consider that more than 90 percent of all your brain synapses are primed to release glutamate. That makes for a lot of glutamate action going on.

At the same time, glutamate is also used by the brain to produce its alter-ego, gamma-Aminobutyric acid (GABA), the major inhibitory neurotransmitter of our nervous system. So not only does glutamate prompt you to get that slice of pizza—it can also stop you from having that slice in the first place.

But it doesn't end here. As it turns out, glutamate features prominently in yet another crucial area of brain health. It also assists in learning and memory.

Here's the story. The connections between two or more neurons are strengthened or weakened depending on how often those neurons are activated in tandem. According to the theory of Hebb (one of the most famous theories in the field of neuroscience), "neurons that fire together, wire together," while those that fire out of sync . . . lose their link. It is widely believed that this process, called *long-term potentiation* (LTP), is based on glutamate, or a slightly different version of glutamate that goes by the rather impressive name of *N-methyl-D-aspartate*, or NMDA for short. NMDA possesses its own specific set of receptors in the memory centers of the brain. These receptors act as gates, which are generally kept locked. Glutamate (in the form of NMDA) is the key to opening these locks. When the neurotransmitter arrives, the gates open, allowing information to flow into the neuron. Over time, the more frequently this happens, the longer the gates will stay open. The feedback loop that ensues forms the basis of *synaptic plasticity*, which is the biological equivalent of memory formation.

But let's put all this information into a brain-food context. Our ability to initiate an action, to refrain from doing something, *and* to form long-term memories are all dependent on the amino acid glutamate.

Glutamate (or *glutamic acid*) is a non-essential amino acid, which means that the brain is capable of producing it on its own. However, once again, there is a catch. The brain needs the sugar *glucose* to make glutamate. Glutamate is formed when the brain breaks down glucose in a process called *metabolism*, which is the very same process the brain uses to make energy. This makes most of our mental activities highly dependent on our dietary choices, and specifically on our intake of carbohydrates.

6

Carbs, Sugars, and More Sweet Things

NOT ALL CARBS ARE CREATED EQUAL

As described in the previous chapter, the brain is a particularly talkative organ. Its activity requires a continued fueling of the electrical impulses neurons use to produce the neurotransmitters and communicate with one another. This awe-inspiring process requires a tremendous amount of energy.

One of the very first questions that nutritionists sought to answer in their quest of a healthy diet was "What keeps us running?" *Carbohydrates* figured prominently in the answer.

There are many different forms of carbohydrates, classified according to their chemical composition as well as their capacity to provide energy. There are carbs that provide quick energy, such as simple sugars like honey, and there are complex carbs that require more digestion for their components to be fully absorbed by the body and consequently provide time-released energy, such as whole wheat and brown rice.

It was the quick-energy type of carbohydrates, the simple sugars, that prompted early nutritionists to recognize these nutrients as the main energy source and fuel for the body as a whole. But of all organs in the body, the brain is the one that needs them above all.

This is yet another major difference between the brain and the body. While the body can use both fat and sugar for energy, the brain relies exclusively on a sugar called *glucose.* In other words, all the energy our hungry brains need—every ounce of it—comes exclusively from glucose. Before sounding the alarm ("Sugar!"), consider that there is nothing exceptional about this. Normally, human bodies are sugar-driven machines: glucose is the go-to nutrient and the quickest way for the entire body to obtain energy. Whenever you eat foods that are naturally rich in carbohydrates, these are ultimately broken down into glucose, which is quickly absorbed in the bloodstream and transported all over the body to be immediately used for energy through the process of metabolism. Glucose readily crosses the blood-brain barrier to feed all those voracious billions of cells that populate your brain.

So don't let the statistics mislead you: while it is true that carbohydrates only account for a very small percentage of the brain's physical composition, there is a never-ending 24/7 influx of glucose inside the brain. Since its work is so demanding, glucose is expended at such a rapid-fire rate it simply doesn't have time to slow down and accumulate in tissues.

Where is all this glucose coming from? Our diet, of course.

From a neuro-nutrition perspective, carbohydrates like glucose are far from being the enemy, as they are essential for proper brain activity and cognitive performance. The human brain is so critically dependent on glucose that it even developed sophisticated mechanisms to convert other sugars *into* glucose. For example, *fructose*, a sugar found in most fruit and honey, as well as *lactose,* found in milk and dairy, can both be transformed into glucose as a quick fix whenever our glucose runs too low.

However, if you are about to reach out for a sugary snack, hold on.

When we talk about carbs, we are not talking about cupcakes. Also, we are not talking about stuffing ourselves silly. Although glucose is one of the few nutrients to be granted immediate entry into the brain, its entry is tightly regulated. In accordance with a strict supply-and-demand model, specific "sugar gates" are present in the blood-brain barrier that open when the brain needs glucose and close once enough glucose has been supplied. If your brain is active and needs glucose to work, it will take up as much glucose as it needs from the bloodstream. But if your brain is feeling satisfied and doesn't need more glucose than it has already absorbed, that last bite of pasta or spoonful of gelato won't make your brain work any harder or better—it would only be facing a closed door. It might make you gain a few pounds in other parts of your body, though.

Once inside the brain, whatever little glucose isn't immediately used to make energy is converted into a substance called *glycogen* and stored away for future use. This is an efficient way of saving useful calories and providing your brain with an energy reserve that will keep you going in between meals. However, glycogen stores are minimal. Our reserves would last no more than a day, if needed.

When carbohydrate supply is limited (usually below 50 g/day, the equivalent of three slices of bread), glycogen stores are quickly depleted, leaving our working brains in potential jeopardy. But ingenious as ever, our brains have a Plan B. If carbohydrate supply runs too low, Plan B goes into effect, and the brain turns to the liver to start burning adipose fat and produces a new type of molecule called a *ketone body*. Ketone bodies are the only backup energy source for our brains.

You might have heard of ketone bodies if you've ever been on certain low-carb diets. One in particular is called the *ketogenic* or "keto" diet—and it is a neuro-nutritionist's nightmare. It is high in saturated fat and very low in carbohydrates and fiber, which forces the liver to burn all available sugars before turning to fat to stabilize blood sugar levels. At the same time, burning fat can promote weight loss—and according to some, even mental well-being. We'll talk more about the keto diet in chapter 9. For now, keep in mind that, while it is true that the brain can use ketones in place of glucose, this is *the exception, not the rule.* Burning ketones instead of glucose is the body's providential mechanism reserved for extremes such as starvation. Were the brain able to ask you for nourishment, it would not be asking for ketones. More important, the brain cannot run solely on these molecules. It still requires no less than 30 percent of its energy from glucose.

All in all, the brain runs best on glucose. In fact, it is so vulnerable to sugar deficiencies that any interruption in glucose supply causes a near-immediate failure of brain function, as can be seen from the rapid loss of consciousness caused by severe *hypoglycemia* (very low blood sugar). Especially as we age, we need to ensure that our brains have access to all the glucose needed to maintain functionality and mental sharpness on a daily basis.

DESPERATELY SEEKING GLUCOSE

Carbohydrates are controversial when it comes to dieting. But from the brain's perspective, what differentiates "good carbs" from "bad carbs" is actually the food's specific glucose supply.

Glucose has been the main subject of my research since college. Over the years, I've looked at glucose in every possible way, from blood tests to brain scans. Your brain needs it. No matter how many dietologists, doctors, or journalists tell you that carbohydrates are bad for you, the brain runs on glucose, and glucose is a carbohydrate.

The problem is, when most people say carbs, they think of white food: sugar, bread, pasta, and baked goods. As sugary as these foods might taste, they are not good sources of glucose.

So where can we find this precious sugar?

As you can see in Table 5, foods that we wouldn't have necessarily thought of as sugary, such as onions, turnips, red beets, and rutabaga, turn out to be the best natural sources of glucose. The red beet in particular is "Nature's candy." A small red beet alone contains 31 percent of all the glucose you need for the day. Fruits like kiwi, grapes, raisins, and dates are also excellent, as are honey and maple syrup. Whether we're speaking generally or looking for a brain pick-me-up, these foods are much better natural sources than others because they provide us with our precious glucose while minimizing the total amount of sugar ingested.

Instead, sugary foods like candy, cookies, and even orange juice contain plenty of other sugars but hardly any glucose. For comparison, white table sugar is 100 percent sucrose, a different type of sugar.

Food item	Glucose (g/100 g product)	Total sugar (g/100 g product)	% Glucose
Spring onion	1.4	1.6	88%
Turnips	1.9	2.5	76%
Rutabaga	2.2	3.9	56%
Apricots, dried	20.3	38.9	52%
Kiwifruit	5.0	10.5	48%
Grapes	6.6	16.4	40%
Onion	1.9	5.0	38%
Whole-wheat bread	1.4	3.9	36%
Red beets	4.0	13	31%
Honey	24.6	57.4	30%

Table 5. Top ten glucose-rich natural foods, ranked by % glucose content.

This brings us to the next question. How much glucose do we need?

Believe it or not, you won't find the answer on the Internet. In fact, as of today, there are no dietary requirements for glucose (or for carbohydrates, for that matter). We need to turn to science to find the answer.

The best way to look at brain metabolism is through PET scans. For many years, I have been using PET to study the way the brain burns glucose to produce energy (glucose metabolism) and its relationship to cognitive functions such as memory, attention, and reasoning.

While everybody is more or less familiar with MRI scans, not many people know what a PET scan actually is. Have you ever seen one of those pictures of the brain painted in bright red and yellow or the darker hues of blue and green? Those are PET scans. The brightly colored areas are the more active of the brain, while the darker ones show a lower brain activity. Since the brain uses exclusively glucose from our diet to stay active and produce energy, we're actually looking at the brain burning glucose from our foods.

This procedure involves injecting a small amount of glucose into the bloodstream. The glucose then rapidly enters the brain and flows straight to the most active regions in the brain, since that's where fuel is needed most. But we've done something special with this glucose. It is attached to a unique, radioactive ingredient called *fluorine 18*, which glows as it's deposited into the brain. We then use a brain scanner to detect its light, which, thanks to its varying intensity, demonstrates the degree and location of metabolic activity taking place inside your head.

Scientists have used this method to discover the exact amount of glucose a healthy brain consumes on a daily basis. In technical terms, the brain burns an average of 32 micromoles of glucose per 100 grams of brain tissue per minute. In plain English, this means that to stay active and healthy, a vital adult brain needs about 62 grams of glucose over a twenty-four-hour period. Some brains need a bit more; some need a bit less to function at their optimal level.

Do 62 grams of glucose sound like a lot of sugar?

It isn't. In fact, it is less than 250 calories a day. And more important, it can't be just any sugar. It has to be *glucose*. For example, 3 tablespoons of raw honey will give your brain all the glucose it needs for the day. As a comparison, you'd need to eat 16 pounds of chocolate chip cookies to achieve the same goal.

SUGAR HIGHS AND SUGAR LOWS

In addition to focusing on the glucose content of the food we eat, we also need to pay attention to our total sugar intake over the course of the day. A major downside of the brain's reliance on glucose is that our mental sharpness is highly vulnerable to any drops in blood sugar levels. Therefore, we need to provide adequate amounts of glucose while keeping our blood sugar levels stable, an essential for proper brain function.

Additionally, you don't want your blood sugar to be too high, either. As any diabetic patient knows only too well, blood sugar levels can easily run amok. But for the vast majority of the population, high blood sugar doesn't occur because of a medical condition or because your DNA has a sweet tooth. It all comes down to what you're eating.

This is how it works. Blood sugar levels are regulated by a hormone called *insulin*. The pancreas secretes insulin when there are large amounts of sugar in a meal. Insulin helps cells and tissues take up that sugar for energy, while at the same time removing the sugar from our bloodstream.

People who consume high quantities of sugar, especially refined white sugar, on a regular basis end up overworking their pancreas and exhausting their insulin sensitivity. Once the pancreas is burnt out, blood sugar levels will remain high regardless of the amount of insulin produced. This leads to a medical condition called *insulin resistance*, in which the body produces insulin but isn't able to use it efficiently. When people have insulin resistance, glucose builds up in the blood, altering the body's metabolism to its great detriment.

Insulin resistance has become a frighteningly common condition in the United States. In 2012, the U.S. Department of Health and Human Services estimates that at least 86 million American adults, ages twenty or older, had insulin resistance or pre-diabetes. The prevalence is estimated to rise even more in the future. People with insulin resistance are at high risk for type 2 diabetes, obesity, and heart disease, which have long been recognized as risk factors for dementia. As many as 6 to 8 percent of all dementia cases are attributed to type 2 diabetes, while heart disease and stroke account for another 25 percent of patients. As an additional side effect, insulin resistance is coupled with increased fat accumulation, which only worsens one's metabolic imbalance, with terrible repercussions for the health of our brains.

Moreover, the hippocampus itself (the brain region specialized in memory) can actually experience insulin resistance. Without going into too much detail, insulin resistance can lead to brain inflammation and accelerated free radical production, making it very hard to remember anything at all.

NOT AS SWEET AS IT SEEMS

Have you ever experienced the highs and lows caused when your blood sugar levels jump too high and then come crashing down? Have you noticed how weak and fatigued you feel afterward? That's "the candy bar effect" and it doesn't do you, or your brain, any good.

As mentioned above, high blood sugar levels cause inflammation, insulin resistance, metabolic disorders, and type 2 diabetes—which, in turn, raise one's risk of dementia.

But you don't even have to go as far as actually having diabetes to be in trouble. Research shows that high blood sugar levels can have a damaging effect on the brain, especially as we age. For example, a study tracked blood sugar levels of over two thousand elderly people for seven years, to compare high blood sugar with the possibility of poor cognitive outcomes later in life. While none of the participants had dementia when the study began, many of those with high blood sugar levels developed dementia over the course of the study. The higher the amount of sugar in the blood, the higher the risk of dementia—*even* at glucose levels considered normal in standard glucose tests (<120 mg/dL). In other words, sugar levels that are "tolerable" for the body as a whole might in fact be too high for our delicate brains.

Even more disconcerting evidence comes from brain imaging studies showing that dementia-free elderly with high sugar levels exhibit not only decreased memory performance, but also accelerated brain shrinkage, as shown on MRIs. This correlation was also found in participants without a trace of diabetes, and was particularly dramatic in the memory regions of the brain.

As a society, if we want to preserve our memories and lower our risk of mental deterioration (and diabetes), we urgently need to limit our sugar intake to just the amount and sort the brain really needs. This means focusing on healthy sources of glucose while reducing our intake of those insulin-crashing bad sugars to keep our metabolism strong and steady.

Unfortunately, the typical American plate is filled with processed food and refined grains, and is often accompanied by large (or by the rest of the world's standards, *gigantic*) cups of soda on the side, each one brimming with concentrated sugar. It's just as common to see people snacking on candy bars, fluffy white-flour products, and extra-large, sugar-enriched fancy coffee drinks. According to the nutrient data provided by Starbucks, a tall (which is actually their smallest portion offered) Caramel Chocolate Frappuccino Blended Crème provides 300 calories of goodness. But hidden behind its inviting appearance, you'll find a whopping 48 grams of refined white sugar. If that weren't enough, you have the option to top it all off with an additional hearty dollop of whipped cream. And if you're looking for fiber, look elsewhere.

The bottom line is that we eat far too much of the sort of sugar that makes us fat and not enough of the kind of sugar that makes us smart. For our own good, we owe it to ourselves and our children to correct this.

A good way to keep an eye on sugar intake is to pay attention to a food's *glycemic index*. The glycemic index is a nutrition rating system that ranks foods based on their ability to affect blood sugar levels. If a food metabolizes quickly into sugar, and the sugar is absorbed quickly into the bloodstream, the food gets a high rating because of its ability to cause an unhealthy insulin spike in your body. If a food barely raises your blood sugar, it is assigned a low score instead. Additionally, it's important to look at the *glycemic load* of your food. This is a similar system that ranks foods not only in terms of how quickly they produce a sugar response, but also by the amount of fiber a food contains. The more fiber, the lower the overall food's effects on insulin.

In terms of brain activity, foods that metabolize quickly into sugar and contain very little fiber are the worst you can eat. These include sugary beverages, sweetened fruit juices, baked goods, and candy, as well as white-flour foods such as pasta and pizza.* Instead, complex carbs and starches are more fibrous and difficult for your body to break down, and the slower breakdown results in less of a blood sugar spike. Sweet potatoes or yams (especially eaten with the skin on), fiber-rich fruits like berries and grapefruit, and vegetables such as pumpkin, butternut squash, and carrots are all excellent lower glycemic foods. Other choices such as legumes (lentils, garbanzo beans, and black beans) and whole grains (with their husks still on) also provide you with more stable sugar levels while at the same time being a good source of brain-essential glucose. In other words, if you have a sweet tooth, the trick is to load up on fiber.

From a nutritional perspective, fibers are divided into soluble and insoluble fiber. Soluble fiber, like that found in oatmeal, blueberries,

* For a complete list, see www.health.harvard.edu/diseases-and-conditions/glycemic_index_and_glycemic_load_for_100_foods.

and Brussels sprouts, is the kind of fiber that turns into a gel-like texture as you eat, slowing down your digestion and making you feel fuller longer. Insoluble fiber, like that found in wheat bran and dark leafy greens, does not dissolve during digestion at all, adding bulk to your stool. This in turn helps your digestive tract eliminate waste more quickly. Many whole foods, especially fruits and vegetables, naturally contain *both* soluble and insoluble fiber.

Besides offsetting your insulin response, the amount of fiber in your diet plays a major role in supporting your gut and immune health. Diets that are low in fiber are typically associated with constipation, gastrointestinal (GI) disorders, inflammation, and an increased risk of some cancers like colon cancer. Those countries whose diets are dependent upon frequent fast-food consumption and little or no fresh produce (such as the United States) top the list of low-fiber eaters, consuming as little as 10 to 15 grams of fiber a day, which also correlates with high rates of GI issues in our country. As we'll see in chapter 8, if your gut isn't healthy, your brain will suffer as well, which further underlines the importance of eating fiber for optimal brain health.

In conclusion, to make your brain happy, focus on low-glycemic/high-fiber foods as the main source of carbs in your diet, and indulge in high-glycemic foods only in small amounts and infrequently.

If you, like me, can't go without an occasional treat, do not despair. Some foods that qualify as "treats" still possess an overall low glycemic load, making a better food for you than originally thought. For example, a square of organic dark chocolate (70 percent or higher) has a low glycemic load, which makes it satisfying *without* the sugar rush. Same for popcorn. I invite you to check my website (www.lisamosconi.com) for several glucose-rich, lip-smacking treats that you can indulge in that won't compromise your blood sugar levels nearly as much as a candy bar would.

7

Making Sense of Vitamins and Minerals

HAVE YOU BEEN TAKING YOUR VITAMINS?

Vitamins play an essential role in our brains' activity, growth, and vitality. Although they are not a source of direct energy, vitamins assist the brain in energy production. More specifically, they provide the key the brain needs to unlock the energy stored in the foods we eat and to activate a variety of metabolic processes that would be in danger of failing in their absence.

The importance of vitamins in relationship to optimal health has been understood by healers throughout recorded history and has long been a subject of scientific study. In particular, the discovery of vitamin deficiencies was a stepping-stone toward the recognition that nutrient deficiencies are at the heart of so many diseases once thought unavoidable.

We now know that vitamins are crucial to boosting our immune system, for absorption and elimination of other nutrients, and even more important, to produce neurotransmitters, the chemical messengers of the brain. As a result, several neurological diseases are made worse or even caused by vitamin deficiencies. Lack of vitamin B1 (*thiamine*) is implicated in degeneration of the nervous system (*polyneuropathy*) and in Wernicke-Korsakoff syndrome, a brain disorder that can evolve into dementia. Insufficient levels of vitamins B6 (*pyridoxine*) and B12 (*cobalamin*) also can lead to dementia. Deficiencies of vitamin B9 (*folate*) are known to cause neural tube defects of the fetus, leading to cognitive dysfunctions later in life. And the list goes on.

All these vitamins come from the foods we eat. Most vitamins cannot be manufactured in our brains or bodies at all, and therefore belong to the group of brain-essential nutrients that need to be obtained from our diet. When you eat fresh vegetables or fruits, the vitamins they release in the bloodstream travel all the way up to the brain, where they are received by their dedicated transport systems and are happily welcomed across the blood-brain barrier.

Vitamins are usually broken down into two categories: fat-soluble (i.e., dissolve in fat) and water-soluble (i.e., dissolve in water). They are further defined by chemical names and subtypes such as vitamin B6 or B12.

Fat-soluble vitamins include vitamins A, D, E, and K. An advantage of fat-soluble vitamins is that they can be stored in the fat of our bodily tissues and, as a result, don't need to be continuously replenished. Among these vitamins, two are known for their brain-protective properties: vitamin A (via its precursor *beta-carotene*) and vitamin E. Both these vitamins have *antioxidant* functions that protect brain cells and tissues from the harmful effects of toxins, free radicals, and even pollution. Moreover, vitamin E increases delivery and absorption of oxygen into brain tissue, which is essential for optimal function and metabolic activity.

All other vitamins are water-soluble. Water-soluble vitamins cannot be stored and as such they are required daily in our diets. These include a number of vitamins that are essential for brain function, such as vitamin C, vitamin B12, vitamin B6, folate, and choline. In general, the major value of these vitamins is their ability to "make things happen." They are the go-getters of the brain; their role is to facilitate the action of our neurotransmitters, sometimes even becoming part of a neurotransmitter itself. This is crucial for proper communication between brain cells. Choline is an example of how this all plays out.

CHERISH YOUR MEMORIES WITH CHOLINE

The brain depends on choline (a B vitamin) to manufacture the neurotransmitter *acetylcholine*. Acetylcholine is one of the main brain neurotransmitters, as it is crucial for memory and learning, as well as arousal and reward. If you or a loved one has Alzheimer's, you have probably heard of how the memory loss typical of the disease is associated with a shortage of acetylcholine. Most available drugs for Alzheimer's, like donepezil (Aricept), are aimed at improving the action of this chemical in the brain.

However, anyone can be low in this neurotransmitter because of their diets. Since choline is an essential nutrient that the brain needs but is not able to produce on its own, the production of acetylcholine is limited by how much choline is reaching the brain at any given moment.

Approximately 10 percent of the total amount of choline circulating in the body is produced by our liver. We must rely on our diet to provide the remaining 90 percent needed. For example, eggs are among the richest sources of choline. Eating a five-egg omelette nearly quadruples choline levels within a few hours, making this nutrient readily available to the brain for acetylcholine production. But if you are eating a diet deficient in choline, you will likely cause an acetylcholine deficiency in your brain—affecting your memory as the by-product. Unfortunately, as much as 90 percent of the American population is deficient in choline.

This begs the question: Is it easy to eat enough of this vital brain nutrient?

Not particularly.

According to current dietary guidelines, if you are an adult woman, you and your brain need at least 425 milligrams (mg) of choline daily, while men need about 550 mg daily.

In practical terms, if you are an adult woman, you can get your 425 mg of choline by eating 22 grapefruits, or 3 pounds of broccoli, or half a chicken . . . or you can eat 3 eggs. If you are an adult man, you'll need more of the same foods, so 27 grapefruits, 4 pounds of broccoli, 2 pounds of chicken, or about 4 eggs. Every day.

This is not to suggest that you should eat several pounds of broccoli a day (nor a case of grapefruit), but it does serve to clarify two very interesting things. First of all, it makes it clear that some foods are more choline-dense than others, and therefore more brain-supportive. For example, it's easier to enjoy a three-egg omelette than to ingest twenty-two grapefruits. Second, this shows that eating right for your brain is both literally and figuratively "not a piece of cake."

Let's now take a look at the foods that best provide the choline necessary to boost our brains' acetylcholine levels. As you can see in Table 6, egg yolks top the list, with a whopping 682 mg of choline per 100 grams of yolk (or about 4 eggs). Since stuffing yourself with egg yolks might be a tad impractical, a combination of different choline-rich foods is arguably the best solution. Other choline-rich foods include fish eggs (caviar), most fish, organ meat (e.g., liver, kidney, brain, and heart), shiitake mushrooms, wheat germ, quinoa, peanuts, and almonds.

Food item	Unit (imperial)	Unit (metric)	Choline (mg)	Choline density (mg/100 g product)
Egg yolk, raw	1	20 g	136	682
Fish caviar	1 tablespoon	16 g	79	491
Brewer's yeast	2 tablespoons	30 g	120	400
Raw beef liver	5 ounces	142 g	473	333
Shiitake mushrooms	1 ounce	28 g	57	202
Wheat germ	1 cup	240 g	202	84
Codfish	0.5 pound	227 g	190	84
Quinoa, raw	1 cup	170 g	119	70
Chicken	0.5 pound	227 g	150	66

Table 6. Top ten food sources of choline. Food items are ranked by choline density.

As an additional assist, nutritional supplements can be helpful. Here's a shortcut that might do the trick. Brewer's yeast (the one that's used to make beer, not to bake cakes!) is a great natural source of choline. Though an acquired taste for some, Marmite, a product that features brewer's yeast as its main ingredient, is a regular staple in most English and Australian kitchens. Since first being manufactured in the early 1900s, this savory spread has been praised for its high nutritional value, so much so that it was included in every soldier's rations during World War I. Considering it takes only a couple of tablespoons of brewer's yeast to reach our daily choline requirements, sprinkling this food on cooked vegetables and salads, or adding to soups and stews, is a smart way to boost your choline levels. I always add some to my soups.

One word of caution: although it's quite difficult to eat too many foods rich in choline, keep in mind that excessive choline can be toxic. In general, no more than 3500 mg for adult men and women is recommended as a daily dosage.

VITAMIN B6: THE PASSE-PARTOUT TO BRAIN ACTIVITY

As mentioned in chapter 5, the brain wouldn't be able to produce neurotransmitters such as serotonin, dopamine, or GABA without the help of vitamin B6, so keep your B6 vitamin intake front and center. This vitamin must be obtained from our diet on a daily basis.

Vitamin B6 is widely available in many natural foods. As shown in Table 7, the best natural sources include sunflower seeds and pistachios, as well as fish (especially tuna), shellfish, chicken, turkey, lean beef, and organ meat. Other good sources can be found in sweet potatoes, avocado, leafy green vegetables, cabbage, bananas, and whole-grain products such as wheat bran and germ. Among all veggies, garlic is surprisingly rich in B6. A little more than 100 grams of garlic would provide all necessary B6 intake for the day. Unfortunately, 100 grams of garlic is equal to 40 fresh garlic cloves, which would be a challenge to ingest, not to mention what it would do to your breath. Vampires or not, that's too much garlic.

Royal jelly (a more potent version of honey) is also a rich source. Personally, I take a teaspoon of royal jelly complete with bee pollen almost daily for their known (though perhaps not scientifically confirmed) natural antibiotic effects. Try drizzling it over yogurt, and top it all with chia seeds and crushed pistachios for a perfect serotonin-boosting snack.

Vitamin B6 supplements are also readily available in over-the-counter capsules or tablets. While toxicity is unlikely to occur from natural food sources, it is recommended that our daily intake of B6 not exceed 100 mg/day for adult men and women. So when using supplements, always keep an eye on the dosages.

Food item	Unit (imperial)	Unit (metric)	Vitamin B6 (mg)	Vitamin B6 density (mg/100 g food)
Pistachios	1 cup	123 g	2.1	1.70
Garlic	6 cloves	20 g	0.22	1.10
Tuna	4 oz	113 g	1.18	1.04
Turkey	4 oz	113 g	0.92	0.81
Beef	4 oz	113 g	0.74	0.65
Chicken	4 oz	113 g	0.68	0.60
Salmon	4 oz	113 g	0.64	0.57
Royal jelly	1 teaspoon	5 g	0.05	0.5
Spinach	1 cup	90 g	0.44	0.49
Cabbage	1 cup	90 g	0.34	0.38

Table 7. Top ten food sources of vitamin B6, ranked by vitamin B6 density.

MORE B VITAMINS: WHAT PROTECTS THE HEART PROTECTS THE BRAIN

Now let's turn our attention to a substance called *homocysteine.* As doctors have long known, high homocysteine (*hyperhomocysteinaemia*) is a strong risk factor for stroke, which in turn is a major risk factor for dementia, accounting for as many as 25 percent of all cases.

Worse still, research shows that high homocysteine levels affect cognitive function even in people *without* strokes. Typically, in a lab test, homocysteine in the range of 4–17 mmol/L is considered safe. However, a study of over a thousand cognitively intact elderly who were followed for several years found that the risk of developing dementia was nearly doubled for those who had homocysteine levels of just 14 mmol/L at the study's start. Even more stunning, an increase of only 5 mmol/L in homocysteine levels raised the risk of cognitive deterioration by another 40 percent. This indicates that our brains are more sensitive to this substance, and therefore to vascular changes, than previously imagined.

The good news is that high homocysteine levels are completely reversible. Not only that, one can achieve this simply by eating right. How?

The production of homocysteine happens to be regulated by specific B vitamins: B12 and folate (i.e., Vitamin B9) above all, in combination with the B6 mentioned earlier. When you don't have enough of these B vitamins in your system, your homocysteine goes up and affects your circulation. When you do have enough, it goes right back down where it belongs.

A number of studies agree that adequate B-vitamin levels are protective against age-related cognitive decline. For instance, a study of one thousand people ages sixty-five and older found that those whose diets were rich in folate (more than 400 mcg/day) had a lower risk of developing dementia as compared to those whose diets were poor in folate (less than 300 mcg/day).

Similar results were observed for vitamin B12. Among over five hundred elderly participants, those who had low B12 levels (i.e., less than the DRI recommended dose of 2.4 mcg/day) had an increased risk of developing dementia as they got older. However, not even those whose B12 consumption was within normal limits were safe. The rate of cognitive decline for the average eighty-year-old who consumed the recommended dose of B12 was 25 percent higher than that of a similar eighty-year-old who consumed 20 mcg/day. Note that this is *ten times* the recommended dosage. Clearly, your brain needs more B vitamins than the rest of you.

It is very important to make sure that our brains have access to plenty of these B vitamins. Not only are they good for your brain at large, but they also have a firmly established role in the prevention of dementia. Recent randomized, double-blind, placebo-controlled trials (that all add up to very thorough trials) tested the effects of high-dose B-vitamin supplementation in a group of 85 patients with mild cognitive impairment (MCI), a condition at high risk for progressing to Alzheimer's. Over a two-year period, the patients were treated with a combination of folic acid (0.8 mg/day), vitamin B12 (0.5 mg/day), and vitamin B6 (20 mg/day). At the end of the study, supplementation with these three vitamins had maintained memory performance and at the same time reduced the rate of brain shrinkage when measured via an MRI. The treatment was particularly effective for participants with high homocysteine levels. Their homocysteine went down to normal levels, and they also showed 53 percent *reduced* brain shrinkage.

Interestingly, the treatment's success was also related to the patients' consumption of omega-3 PUFAs. Those with high levels of omega-3s responded extremely well to B-vitamin supplementation. On the other hand, patients with low omega-3 levels before and during treatment did not show improvements, but rather showed the same brain shrinkage as the untreated patients. The moral of the story: make sure you consume plenty of omega-3s together with your B vitamins to maximize their combined efficacy.

These B vitamins are easy to obtain from a balanced diet. Many plant-based foods are rich sources of folate, especially black-eyed peas, lentils, spinach, tofu, and avocado. Vitamin B12 is found in shellfish (clams are the richest source) and fish (salmon, trout, mackerel, and fresh tuna fish top the list), as well as chicken, eggs, beef, and dairy products. Remember how a mere 3 ounces of wild salmon supply enough omega-3s for an entire day? Well, that same small piece of fish also contains three times the recommended daily dose of B12. Pair that with a fresh spinach salad and some avocado, and you've got a perfect brain-healthy meal.

In the end, just imagine that 25 percent of all dementia cases, and most likely some of the remaining 75 percent, might be fully prevented by increasing consumption of B-vitamin-rich foods. Ditto for strokes and vascular disease. It is hard to believe that after all this, some people still are not convinced that food is medicine.

THE BRAIN'S DEFENSE SYSTEM: THE ANTIOXIDANTS

Have you ever left an apple cut in two on the counter and noticed how it begins to brown? Were you to leave it there, it would darken further and begin to shrivel up. If it were in sunlight, this process would quicken even more—and should the environment be a polluted one, the entire process would accelerate further still. This sort of "rusting" effect is what we refer to as *oxidation*. Oxidation itself is a very normal process. It happens all the time to our brains, bodies, and many things that surround us—like apples left on the counter, or iron pipes left in the rain.

In the brain, oxidation happens whenever our brain cells burn glucose and oxygen to produce energy. And as you might guess, this is happening continuously. Generally, our brains manage to counterbalance this ongoing oxidation, but sometimes things get out of hand, and the amount of oxidation exceeds our brains' ability to keep it in check. In that case, our brains suffer from what's called *oxidative stress*. Simply put, oxidative stress is the damage made to cells as a result of prolonged oxidation and of the action of free radicals—harmful molecules that are also produced in the process.

Of all bodily organs, the brain is the one that suffers most from oxidative stress. Free radicals incessantly develop, thought by thought, making their way through our neurons like little tornados. The more free radicals your brain contains, the more damage done.

However, we are not defenseless. We can call upon *antioxidants* to protect our delicate brains. These substances are Nature's way to prevent oxidation from happening in the first place. They have the capacity to wander throughout our bodies (brain included), fighting off whatever free radicals they encounter along the way. In simple terms, antioxidants are our police officers chasing away the bad guys.

Some antioxidants are produced by our body, but most are not and need to be obtained from our diet. In particular, vitamin E (from almonds or flaxseeds) and vitamin C (from citrus, berries, and a variety of veggies) are the body's main antioxidant defenders. As an experiment, try sprinkling some lemon juice on that shriveled apple and watch how much longer the apple lasts.

It is important that we eat enough antioxidants to protect ourselves against brain aging and disease. Large-scale studies in the United States and Europe found that those elderly who consumed at least 11 IU (16 mg) of vitamin E per day had a 67 percent lower risk of developing dementia as they got older compared to those who consumed 6 IU (4 mg) per day. Those who consumed the highest amounts of both vitamin C and E had an even lower risk. Studies estimated that we need 133 mg or more of vitamin C per day to provide our brains with optimal protection in addition to the 16 mg of vitamin E specified above.

In more practical terms, it just so happens that the spinach-avocado salad mentioned in the previous section is rich in B vitamins as well as vitamin E. Sprinkle it with lemon juice for extra vitamin C, *et voilà*. See how easy it is to eat right for your brain?

Overall, there is consensus that regular consumption of vitamins C and E, sometimes along with beta-carotene (a precursor to vitamin A found specifically in fruits and vegetables orange in color), reduces the speed at which our brain cells age, increasing longevity and lowering the risk of cardiovascular disease and dementia.

Yet current guidelines recommend we take much lower doses of these precious nutrients. Part of the problem is that, when tested in formal clinical settings, antioxidant vitamins didn't turn out to be as miraculous as everyone had hoped. Vitamin E was the only one that showed potential for slowing functional decline in Alzheimer's patients—but only at high doses of 2000 IU (1.3 grams) per day.

What was at first puzzling to many led to a revelation: antioxidant *supplements* don't really work. We must obtain these nutrients from natural food sources instead. The truth is, in the studies mentioned, only those participants who obtained their antioxidant vitamins from *foods* showed lower rates of cognitive decline and dementia. Those who relied on supplements to get their antioxidants had the same chance of developing dementia as those who consumed little to no vitamins.

Vitamin E is a great example of why this happens. The vitamin E found in synthetic supplements is composed of only one of the eight natural forms of this vitamin (alpha-tocopherol), while food itself provides combinations of all forms. This seems to reduce oxidative stress and inflammation to a greater degree than alpha-tocopherol alone. Pair that discovery with the fact that pills aren't a particularly delicious meal, and it only makes sense to consume your antioxidants from fresh, vibrant foods such as high-quality vegetables and fruits, as well as nuts and seeds.

THE WONDROUS WORLD OF FLAVONOIDS

As you might remember from chapter 1, plants produce a vast array of chemical compounds called *phytonutrients*. Sometimes referred to as "vitamin P," these substances serve a very specific purpose. Their job is to fight oxidative stress and inflammation, thereby increasing the life of the plant itself. They are usually produced in combination with various antioxidant vitamins and are particularly concentrated in the berries of the plant. As it turns out, plants aren't the only ones benefiting from these compounds. By eating these berries, we, too, receive all the benefits of the plant's effort to "live long and prosper."

Scientists have identified and categorized more than four thousand phytonutrients, such as flavonoids and phenols. Common examples are the *quercetin* found in apples, the *flavanols* found in cacao beans, and the *resveratrol* that gives red wine its good name—each renowned for their superior anti-aging properties. While scientists have historically overlooked these substances, new experimental studies suggest that phytonutrients play a greater role in human health than previously thought. In the next chapters, we'll see how they make up the heart and soul of those diets proven to promote health and longevity worldwide. Meanwhile, let's talk about minerals.

MINERALS: A LITTLE GOES A LONG WAY

Besides vitamins, the brain loves mineral matter—earth-derived minerals that we primarily absorb from fruits and vegetables. Minerals are in fact remains of plant and animal tissue contained in the soil. "Ashes to ashes, dust to dust" refers to this very process. Minerals contained in the soil are absorbed by plants as they grow during the natural process that recycles these elements as nutrients in our foods.

Much like vitamins, minerals are essential to our physical and mental health. They lend structure to our cells, particularly our blood, nerve, and muscle cells, as well as to those that form our bones, teeth, and soft tissues. Minerals also serve many functions specific to the brain. Some function as electrolytes to help regulate brain fluids and hydration. Others power our metabolism. Others yet have the very important task of regulating nerve transmission. Magnesium, zinc, copper, iron, iodine, selenium, manganese, and potassium are each essential in keeping our brains healthy and active as we age.

But it isn't only mineral deficiencies that are harmful to the brain. An excess of some minerals, especially in the form of metals, can be toxic to the brain when absorbed in high concentrations. These are primarily lead, cadmium, and mercury, which are known as heavy metals. Poisoning can easily occur as a result of industrial exposure, air or water pollution, foods, medicines, improperly coated food containers, or the ingestion of lead-based paints. Arsenic, another toxic metal, is used in the manufacture of pesticides and given as an antibiotic to farm-raised animals like chicken (which is unfortunate, as we end up ingesting the poison when we eat the chicken). Nickel is used in the manufacturing of the trans-saturated fats used to make margarine and soft spreads. You might want to check your toothpaste, too. There's a good chance it contains titanium. A seemingly harmless material like aluminum also poses a major threat to our delicate brains. It's long been known that aluminum is toxic to brain cells, even in small amounts. Yet we are all easily exposed to this metal by using aluminum containers and several types of cosmetics and medicines, and even by drinking purified water.

While symptoms and physical findings vary according to the metal accumulated, metal poisoning affects the activity of cells in the entire nervous system, in severe cases leading to brain inflammation (*encephalopathy*), which is often irreversible. Heavy metal toxicity is therefore one of the biggest threats to brain health in our society because of the unrestricted use of these elements for industrial purposes as well as a generalized indifference to the way we treat the planet. However, other more common and far less dangerous minerals can also damage the brain if ingested in excessive amounts. These are iron, copper, and zinc.

In adequate concentrations, these minerals are essential for optimal brain function. Iron is crucial for production of hemoglobin (the part of our blood cells that carry oxygen) as well as of some proteins. Copper is key to enzyme function and the health of our immune system, blood vessels, nerves, and bones. Zinc is one of the most important metals when it comes to supporting brain metabolism. We can all imagine just how easily deficiencies in any of these minerals would affect our brainpower and mental sharpness.

A common medical condition called *anemia* is a good example. Anemia develops when your blood lacks enough healthy red blood cells or hemoglobin. This is typically due to iron deficiency. Some of the first symptoms of anemia are fatigue, dizziness, weakness, loss of stamina, and brain fog, which would have an obvious impact on our ability to perform physically and intellectually.

Fortunately for us, it takes very little iron to reach adequate levels. On the other hand, most people err in the opposite direction, ingesting too much of this nutrient, which is toxic in excess. Make note: too little iron will make you anemic, but too much can harm your brain. This goes for copper and zinc as well.

Some studies have suggested that overconsuming iron, zinc, and copper might contribute to cognitive problems in the elderly. Excessive intake of these minerals promotes oxidative stress, that sort of "rusting" effect that makes your brain age faster.

Copper is the one that seems to be particularly bad news for your brain. New research suggests that the copper we ingest just by eating the typical modern diet is enough to increase our chances of developing Alzheimer's. Copper seems to reduce the brain's ability to clear away toxic amyloid proteins before they form the plaques that are the hallmark of Alzheimer's while encouraging the clumping of those proteins at the same time.

Yet this is nothing compared to the copper ingested as part of a high-fat diet. Research shows that people whose diets are high in copper, saturated fat, and trans fat have particularly quickening rates of cognitive decline—roughly *nineteen additional years' worth of aging.* In these studies, the copper intake that was ultimately associated with cognitive decline turned out to be harmful only among those people whose diets were high in saturated fat and trans fats at the same time. To make things worse, it takes very little copper for brains on a high-fat diet to go haywire—only 2.7 mg per day, which is the equivalent of a mere 3 ounces of ham.

But before we blame food in general for these effects, consider this. What is unknown to most of us is that copper also enters the body via other sources—by way of our drinking water, for one, when delivered via copper pipes. Further, many people absorb extra copper via vitamin supplements. Common multivitamins contain both copper and iron, sometimes even exceeding the recommended dosage. In short, if your diet is high in fat, you really need to watch your mineral intake, too, especially if it comes in the form of multivitamin supplements.

In the United States, we don't even require these particular minerals in our supplements, as we consume plenty of these minerals from our everyday foods.

8

Food Is Information

GENES LOAD THE GUN.
LIFESTYLE PULLS THE TRIGGER.

The past chapters have revealed just how much our brains are affected by our diet and lifestyle choices. From the way our neurons communicate with one another to the birth and growth of new brain cells, our personal day-to-day choices are continually influencing what's going on inside our brains. We will now see how these effects are controlled not only by our lifestyle choices but also by the interplay of our behavior *and* genetic predisposition. Let's begin by addressing the delicate and highly complex theme of our *genetic individuality.*

As Dr. James D. Watson said back in the 1960s, "We're not all equal, it's simply not true. That isn't science." As the quote points out, it takes none other than the man who discovered DNA's double helical structure to confirm just how complex, variable, and highly individual our genes are. This inherent individuality occurs in a substantial portion of our genome, giving rise to distinctive characteristics such as hair and eye color. These variations depend on the subtler genetic information carried in our DNA. Whenever you are looking into a loved one's eyes, you are experiencing nothing less than a distinctive display of their signature genetics at work.

How this actually occurs is less obvious.

Our genetic variability is the result of thousands of years of continuous genetic mutations. A mutation is a permanent change in our DNA. For example, originally, all humans had brown eyes. But some six thousand to ten thousand years ago, a genetic mutation occurred that produced the first blue-eyed human. When this first happened, one can only imagine the sensation it caused! Since then, the blue-eye gene has spread the world over and today is considered a fairly common trait.

Genetic mutations have happened throughout the course of evolution. Some are positive, such as the mutations that ultimately led to the beauty of different eye colors or to the increased size and power of our brains. Other mutations are harmful and lead to disease. However, these "bad" genetic mutations are rare, affecting less than 1 percent of the population.

To sum it up, there are a handful of genetic mutations that can make us sick, and a myriad of genetic variants that simply make us different.

This is particularly true as far as our brains are concerned. The human genome sports an estimated 15 million variations, a large part of which involve brain function. Just imagine that each and every one of us carries at least some permutations of all these variations, and the term "diversity" assumes a whole new meaning.

Our brains possess something akin to a fingerprint. While the architecture of the brain, with its various partitions into functional areas and specific structures, might be roughly the same in all of us, there are large variations when it comes to brain size, shape, activity, and molecular composition. Such differences are not only based on our unique genetic makeup, but are shaped, molded, and written upon by our backgrounds, education, and experiences. Add to that the many foods we've been exposed to, our cultural environments, and all the places we've explored since the day we were born, and it only makes sense that no two brains could ever be alike.

This tremendous variability is never more evident than when viewing brain scans. I have been doing brain imaging for over fifteen years, inspecting and quantifying hundreds, if not thousands, of scans. Among them are those of the young and old and every age in between, those of men and women, those of the happy and the unhappy, of the healthy and less fortunate. In addition, I've studied equal numbers of scans showing the effects of neurological diseases like Alzheimer's, Parkinson's, or stroke. Not a single day goes by that I do not stand in awe at the uniqueness that each patient's scan reveals, each one different and distinct from the next.

In the end, it is our unique genetic makeup, combined with our own lifestyle and behavior, that determines the fate of our brains, and therefore our chances of aging gracefully over the course of a lifetime—or of forgetting names and faces instead.

While it is true that one's brain blueprint largely depends on the DNA it has received from its parents, recent discoveries have led to rethinking the old view that "you are your DNA" in favor of a much more dynamic model. In this new model, genes are pivotal in establishing some aspects of brain health, but it is our current, ongoing lifestyle choices that play a central role in *turning those genes on or off*. As strange as this might sound, you actually have the power to activate or silence your genes, and this discovery is called *epigenetics*.

Epigenetics refers to the fact that while your lifestyle choices won't modify the structure of your DNA, they do have the ability to modify the way your DNA works. Where you live, who you interact with, how you exercise, which medications you take, and—yes, you guessed it—especially what you eat cause changes inside your body that in turn switch your genes on or off. This can occur once in a lifetime or continuously over time, thereby influencing your chances of retaining or *not* retaining optimal cognitive fitness.

What all this boils down to is that your DNA is *not* your destiny after all. The genetic lottery might determine the cards in your deck, but the way you are living your life deals you the hand you are actually playing. We're back to where we started—genes load the gun, but lifestyle pulls the trigger.

SWITCHING OFF YOUR DNA

Among all the lifestyle factors that are known to have an impact on the action of human DNA, food is the one that plays a predominant role. Everybody exercises once in a while, takes medications now and then, or is exposed to environmental toxins on occasion (which all impact your DNA). But when it comes to food, we partake in meals each and every day, and several times a day at that, and we do this consistently over the course of an entire lifetime. It is this continuous exposure to food that makes diet the most important factor ever to affect our DNA.

Several studies have shown that some dietary nutrients have the ability to influence and regulate our DNA's behavior. These happen to be the same nutrients the brain needs most, such as omega-3s, choline, several antioxidants, and B vitamins. Over the past decade, this realization has prompted nothing short of a revolution in the nutrition field. It has in fact become clear that the effects of nutrition on health can only be fully appreciated with a deeper understanding of how nutrients act at a genetic and molecular level.

The interaction between food and genes has become the major focus of a new discipline called *nutrigenomics*, which aims at revealing how food directly influences DNA activity. This novel perspective has brought new meaning to the old adage "We are what we eat" by demonstrating that what we eat is busy determining what we are to become. At the same time, our genes affect our reaction to food as well, making us receptive to certain foods and intolerant of others.

It turns out that diet, far from merely being a source of fuel or sustenance, is instead a "genetic on/off switch." Some foods directly influence our DNA by turning on the good genes that make us more resistant to disease, while others turn off these same genes, making us more likely to get sick.

This is because food is *information*. Dietary nutrients are nothing less than biological signals that, upon entering our systems, are "read" by your cells. Believe it or not, your cells are coded with detectors that are busy searching for specific nutrients. Let's say they spot healthy omega-3s entering your bloodstream via your meal. Once they do that, they let your DNA know that help is on the way. Then it's as if your DNA takes a deep breath and slows down the body's production of other anti-inflammatory compounds as a result. This is just one example of how a common dietary ingredient can powerfully influence your genes. Depending on whether a nutrient is considered friend or foe, a corresponding genetic response will be cued accordingly.

Whether we look at how food affects our genes or how our genes affect our reaction to food, this new, burgeoning science has thrown open the doors to our genetic individuality being key when it comes to how we approach health and nutrition. As a result, the age of a one-size-fits-all approach is fast becoming a thing of the past. Instead, we stand at the forefront of a new, personalized approach to our health, and in turn, our nutrition.

BIO-INDIVIDUALITY

These discoveries point out that there is not one perfect way of eating that will work equally well for everybody. This concept, often referred to as *biological individuality*, is an idea currently gaining unprecedented respect and interest in the medical field.

Bio-individuality insists on the fact that each human being has a unique biochemistry that influences behavior, mental health, hormonal production, allergic tendencies, immune capacity, and of course, nutritional needs. Because of the genetic differences in the ways our bodies process food, some of us are naturally deficient in some nutrients while possessing an overabundance of others. As the proverb goes, "One man's meat is another man's poison." Indeed, it was recently discovered that many human genes have a heightened sensitivity to diet.

This is true both for the individual and for certain populations. Lactose intolerance is a perfect example. Many people are lactose intolerant, which means that they have difficulty digesting *lactose*, a type of sugar contained in milk. This happens because they lack an enzyme called *lactase*, which is responsible for breaking down lactose. Since humans digest mother's milk as infants, the gene that produces lactase switches off on its own post-weaning.

However, once humans began herding cattle, being able to digest cow's milk as adults became an evolutionary advantage. Milk is a good source of fat and protein, as well as calcium, vitamin D, and several B vitamins including the brain's beloved choline. Some populations adapted to this by keeping the lactase gene turned on and thus continuing to produce lactase in their bodies throughout adulthood. Other populations that were not dependent on cattle (such as some parts of China, Thailand, and Africa) did not develop this ability. To this day, people who stop producing lactase after weaning are lactose intolerant as adults.

At an individual level, we find even more varied and unpredictable differences. As a result, we are each genetically unique in the way we process our food. For example, some people are naturally less efficient at assimilating brain-essential nutrients such as vitamin E, some B vitamins, or omega-3s. Others have difficulties controlling trace minerals like copper, iron, and zinc. Still others have insufficient levels of stomach acid or impaired intestinal function and have trouble digesting their food. Making things more complex yet is the fact that we each have a different *microbiome.*

MEET YOUR MIGHTY MICROBIOME

The term "microbiome" refers to the collection of bacteria, viruses, fungi, and other microbes that inhabit the human body.

Just like planet Earth has its own ecosystem filled with animals, plants, and all sorts of organisms, the human body hosts a complex ecosystem of its own, which is home to a fantastic diversity of life. Of this ecosystem, very little of it actually belongs to our species. All sorts of microbes can be found thriving upon our skin, inside our mouths, and along all the body's various nooks and crannies. They have a particularly large stakeout in the GI system. The body of an adult human harbors nearly 100 trillion bacteria, with more than 95 percent of them located in our gut.

Scientists have long known that bacteria reside within humans, but their presence and relevance remained underappreciated until the discovery of the microbiome. It turns out that in our bodies, bacterial cells outnumber human cells by about 10 to 1. In other words, up to 90 percent of our cells are *non-human*. Even though bacterical cells are much smaller than ours, ninety percent is a lot of cells. If you grouped them together they would be about the size of a football and weigh up to 3 pounds.

Not only are there millions of bacteria residing inside us, but these bacteria derive from thousands of different species, each equipped with their own genetic material. As a result, we are literally inundated by non-human DNA. It is nothing short of mind-blowing that the human genome (aka our DNA) is extremely small as compared with that of these much simpler organisms. Our DNA comprises about 23,000 genes, while the microbes inhabiting our bodies are representative of an incredible 4 million distinct genes of their own.

This raises a whole range of thought-provoking existential and scientific questions, not the least of which is: Should we be worried?

Some microbes, especially viruses, can definitely harm humans. For example, some viruses give us measles and the flu. Some bacteria can also be detrimental, like when they cause strep throat or food poisoning. But in reality, fewer than 1 percent of bacteria causes disease in humans. The vast majority are not only harmless but downright helpful. As it turns out, our gut microbes are major players in our overall health, literally from head to toe.

First and foremost, they help us digest food, each one having the ability to absorb various nutrients to a greater or lesser degree, and each one reacting differently to the foods we eat. For instance, our capacity to absorb vitamins like B12, along with several minerals essential for a healthy nervous system, is highly dependent on the health and diversity of our gut microbiome. Moreover, these friendly bacteria are able to produce essential vitamins like folate and also help maintain adequate levels of amino acids like tryptophan, which is in turn necessary to produce neurotransmitters like serotonin.

Additionally, our gut flora produces fatty acids that are beneficial for the body, like *butyrate*, an excellent source of energy for our muscles. But what is particularly surprising about these bacteria-made fatty acids is that they can directly alter the function of the blood-brain barrier, the cellular fortress protecting the brain against infections and unwanted pathogens. The fatty acids produced by our gut microbes can both strengthen the barrier and make it more lax, effectively regulating how many nutrients and foreign substances can pass through the brain's defense system.

Last but not least, our gut flora acts as a mighty warrior for the immune system, defending us against disease-causing microbes. There is a delicate balance between the gut's strength in keeping out the bad guys while at the same time managing to absorb and harbor the good guys. In general, the intestinal lining must be permeable enough for nutrients and other molecules to both enter and exit the intestine. However, if the lining becomes too permeable, "leaky gut" can occur. In this case, the space between cells opens up too much, allowing leakage of intestinal contents, such as large food molecules or bacteria, into the circulation. When our body senses these foreign invaders in our bloodstream, it triggers an inflammatory response, one aimed at escorting the intruders back out. In the long run, the inflammatory response aimed at defending us from the bad guys can backfire, damaging our intestinal cells and microbiome, too, making the gut even more inflamed and leaky. This vicious circle further impairs the system's ability to absorb the proper nutrients, leaving it vulnerable to food sensitivities or allergies. If this weren't disturbing enough, as we'll see in the next pages, the brain is potentially at risk of being affected as well.

BUGS AND BRAINS

Recent studies have revealed that alterations in the gut microbes can influence the risk of brain disorders such as autism, anxiety, depression, and even dementia. This has resulted in increased attention to the concept that a healthy gut is critical to a healthy brain.

I can only tell you the beginning of this story because the story has only just begun. While there's been an explosion of research on the microbiome with regard to brain health in recent years, this field is still in its infancy. It is also important to underline that most microbiome research so far (including antibiotic treatment or even fecal transplants) has been largely based on experiments using rodents. Given the huge differences between mice and men, there is no guarantee that these findings will hold up in humans. That said, some initial studies done on humans support the existence of a relationship between the world of the microbiota and the health of our brains. This initial research has triggered tremendous interest by the professional and lay media, not to mention national funding agencies, leading to a justifiably meaningful shift in the way we view many brain diseases.

Historically, the gut and its collection of microbes have been largely ignored by Western neurology and psychiatry. To this day, students are taught that the brain is anatomically isolated and well guarded by the blood-brain barrier that keeps out pathogens *including* bacteria. There are some exceptions, like when such pathogens happen to bypass the blood-brain barrier causing disease, as in the case of *meningitis*. But for decades, scientists thought of microbes as either fairly harmless things hitching a free ride through our bellies (which therefore had nothing to do with what was happening inside our brains) or direct threats to our well-being that had to be gotten rid of.

This view has dramatically changed, as a number of new studies have shown that our gut bacteria not only influence how people eat but also how we think and feel.

Some of the most convincing work in this regard has been done on anxiety and depression. For instance, animals genetically engineered to be *without a microbiome* (the so-called germ-free mice) have increased anxiety-like behaviors and an exaggerated response to stress. They also demonstrate bizarre behaviors, tend to be antisocial, exhibit memory problems, and even show reckless tendencies. However, scientists found that they could stabilize the animals' behavior by supplying them with friendly bacteria. This not only lowered their stress levels but also directly increased production of GABA (the neurotransmitter that calms nervous activity) in their brains.

Additionally, the microbiome turned out to be closely involved with neuro-development. For decades, doctors and parents alike have noted that anywhere between 40 percent and 90 percent of children with autism also demonstrate some GI symptoms, such as food allergies and digestive issues. Recent studies are showing that there might indeed be a connection with issues present in the child's microbiome. For instance, some of the symptoms expressed by germ-free mice, such as limited social interactions, a tendency toward repetitive behavior, and a reduced communication with their peers, are similar to human autism. Treating these "autistic" mice with *Bacteroides fragilis* (a bacterium that is sometimes reduced in children with autism) improved the animals' behavior. It made them less anxious, more communicative with other mice, and less prone to showing repetitive patterns.

As you might have noticed, all the research reviewed so far was carried out on lab animals. However, there is some evidence that *probiotics* (i.e., live microbes that reside in your gut and benefit the host, aka you) can alter brain function in humans as well.

One of the best-known studies to date used functional MRI (fMRI) to test whether eating probiotic foods like yogurt would elicit changes in brain activity in a group of young participants. Functional MRI is a brain imaging technique that measures changes in blood oxygen levels as a proxy for neuronal activity. This method gives us a snapshot of those brain regions that get activated (or not) in response to different forms of stimulation. In this study, twenty-five healthy women were divided into a group who ate a cup of probiotic yogurt twice a day for a month and a control group who did not eat any yogurt. All participants were then shown upsetting pictures of people with angry, sad, or frightened facial expressions to gauge their emotional responses while being monitored with fMRI. Perhaps surprisingly, there were significant differences in the way yogurt eaters and non–yogurt eaters reacted to this test. The former showed a more moderate response in the face of negative emotional stimulation as compared to the latter. In other words, they were able to remain calmer than those who were without the probiotic supplementation. Wouldn't it be great if doctors prescribed yogurt instead of Xanax?

Besides the microbiome's role in influencing anxiety and stress levels, new research indicates that it might be a preeminent factor in determining brain longevity. Over the course of a lifetime, people who consume a diet high in fiber and low in animal fat (animal fat does not agree with our friendly gut flora) have the healthiest microbiomes. On the contrary, the microbiome of those whose diets are low in fiber but high in animal fat are very fragile and tend to easily collapse. These findings suggest that the lessening of healthy gut bacteria could contribute to the cognitive decline observed in old age.

Many people are even wondering whether dementia itself could be due to bacterial infections or malfunction. As of today, there is no clear evidence that an unhealthy microbiome would in and of itself cause dementia. Nonetheless, many viruses and bacteria can profoundly affect the brain, generating symptoms like confusion, brain fog, and memory loss. For example, the *human immunodeficiency virus* (HIV-1) that causes AIDS can also cause a form of dementia known as "HIV-related dementia" with symptoms that mimic those of Alzheimer's. The *herpes simplex virus* that causes sores around the mouth can trigger brain inflammation (*encephalitis*), thus eliciting cognitive and mood disturbances. *Syphilis*, a well-known bacterial disease contracted chiefly by infection during sexual intercourse, can spread to the brain and lead to severe cognitive impairment. Because of their ability to affect cognitive health, these pathogens are routinely screened for in the evaluation of dementia.

When I was at NYU, we had an interesting experience with a woman referred to us with a diagnosis of MCI, often a prodromal stage to Alzheimer's. She was literally terrified at the thought that she might end up developing dementia, as her mother had died of it just a few years prior. After running several blood tests, our medical director spotted something unusual: the patient had a massive yet completely asymptomatic UTI (urinary tract infection). She reported no pain, no irritation or itching—despite the fact that her urine was full of bacteria and blood cells. Needless to say, antibiotic treatment was initiated immediately. When the patient returned to complete her evaluations a few months later, she received a diagnosis of normal cognition. You can just imagine her relief at being herself again, with no Alzheimer's in sight.

In the end, even though the data is still limited, it seems to point to the fact that the microbiome is involved in several aspects of brain health and behavior. This raises hope that optimizing our diet and lifestyle to favor healthy gut bacteria might also be a viable strategy to manage, or even prevent, anxiety, depression, and autism, as well as the cognitive changes that might occur with aging. But how exactly are we to do that?

GOOD HABITS: PREBIOTICS, FIBER, AND FERMENTED FOODS

First and foremost, our gut health relies on regular consumption of both prebiotic and probiotic foods.

Prebiotics are literally food for your body's good microbes. This is because these foods are particularly rich in a distinctive kind of carbohydrate called *oligosaccharides*, which happen to be your gut flora's favorite meal. These carbs are unique in that, while all other carbs are broken down in the small intestine, oligosaccharides can't be digested there and consequently flow down to the large intestine virtually untouched. Here, they play the critical role of feeding our friendly bacteria and keeping them healthy. These bacteria-supportive carbs come from foods that are not particularly sweet but do have an ever-so-slightly sweet aftertaste, such as onions, asparagus, artichokes, and burdock root. You can also find prebiotics in garlic, bananas, oats, and milk.

Besides nourishing our friendly bacteria, certain oligosaccharides are drawing increasing attention because of their cholesterol-lowering, cancer-fighting, and detoxification effects. These include the *beta-glucans* found in mushrooms (reishi and shiitake mushrooms have shone in many studies) and the *glucomannans* abundant in aloe vera juice. I'm a big fan of both, so expect to hear more about these foods in the next chapters.

Additionally, fiber-rich foods are crucial to our microbiome's well-being by supporting digestive health and regularity. A healthy digestion is key to removal of waste products, harmful toxins, and bad bacteria—

all of which can harm our gut flora. Cruciferous veggies like broccoli, high-fiber fruit like berries, and all kinds of leafy greens, as well as legumes and unsugared whole-grain cereals, are good examples of fibrous foods we should consume on a regular basis to ensure proper gut health.

Besides prebiotics and fiber, our gut microbes crave *probiotic* foods. These foods contain live bacteria (probiotics) that, upon reaching the intestine, replenish our microbiome's good guys. Probiotics are naturally provided by fermented and cultured foods, including fermented milk such as yogurt and kefir, but also pickled vegetables like sauerkraut. More specific recommendations are included in chapter 12.

BAD HABITS: ANTIBIOTICS, MEAT, AND PROCESSED FOODS

In addition to knowing what to add to your diet and lifestyle, it's equally important to know what to avoid. Any foods and agents that are disruptive to your gut's health, either by increasing its permeability or by causing inflammation, can also decimate your microbiome in the process.

Antibiotics are the first entry on the microbiome's Most Wanted list. The microbiome is negatively affected by the overuse of antibiotics, since antibiotics are not fussy and inadvertently kill our good microbes along with the bad ones. Up until World War II, when medical conditions like pneumonia and wound infections could often prove fatal, antibiotics were a major success. However, they have since become overprescribed in many countries, leading to a pandemic of antibiotic-resistant disease. At the same time, there is the added complication of their having diminished both our microbiota's stability and its diversity.

I am not by any means suggesting that you should avoid taking antibiotics when you need them. However, many Americans take antibiotics as a quick-fix measure or even "just in case." For example, I have heard people say, "I have the flu, I need antibiotics." Contrary to popular opinion, that's not necessarily the case, as the flu is often caused by viruses, not bacteria. Discuss this with your doctor the next time you feel under the weather. Incidentally, most European doctors recommend eating yogurt (or taking a probiotic supplement) before or with your antibiotics to protect your GI tract and at the same time replenish your bacterial pool.

After medicines, food is the major factor influencing our intestinal function. While antibiotics might enter our bodies only once in a while, food is constantly altering the status and health of our gut microbes. Of all foods known to have a negative impact on our microbiome, commercially raised meats top the list.

Believe it or not, meat can be a major source of deadly "superbugs." Animals raised in confined animal feeding operations (CAFOs)—which is the norm in modern-day factory farming—are routinely given low doses of antibiotics to prevent diseases caused by the crowded and unsanitary conditions in which they are forced to live. In fact, of all the antibiotics sold in the United States, as much as 80 percent of them are used to treat livestock instead of people! The problem is that when we eat the meat, we also ingest the antibiotic. As a result, for many people meat is the primary source of antibiotic overload.

What's worse, half of all the meat sold in American grocery stores harbors drug-resistant bacteria that can cause severe food-borne illnesses. According to a recent study by the U.S. Food and Drug Administration (FDA), antibiotic-resistant strains of *Salmonella* and *Campylobacteron* were found in 81 percent of ground turkey, 69 percent of pork chops, 55 percent of ground beef, and 39 percent of chicken meat all across the country. Even more disconcerting, federal data shows that 87 percent of all meats tested positive for *Enterococcu*s bacteria and *Escherichia coli* (*E. coli*), which means that the meat had at one point come in contact with fecal matter.

This is one of the many reasons why I recommend eating only organic, grass-fed, free-range, or pasture-raised meat, dairy, eggs, and other animal products, since organic standards do not allow non-medical use of antibiotics.

Processed foods are another major threat to our gut. Besides being high in unhealthy sugars, such as high fructose corn syrup and refined white sugar, processed foods often contain *emulsifiers*, which are particularly detrimental to the microbiome. Emulsifiers are food additives used to improve the texture, appearance, and shelf life of many foods, and are included in everything from ice creams to baked goods, salad dressings, creamers, and dairy and nondairy milks (yes, even your "healthy" almond milk can be bad for you if it contains emulsifiers). It turns out that these substances can increase permeability in the gut lining, causing an influx of bad bacteria into the bloodstream. This in turn can cause colitis and intestinal inflammation such as irritable bowel syndrome (IBS), as well as the metabolic dysfunctions that can lead to obesity, high blood sugar, and insulin resistance.

Next time you go grocery shopping, take a closer look at the ingredients label on any packaged foods you pick up and check for these common food additives: lecithin, polysorbates, polyglycerols, carboxymethylcellulose, carrageenans, xanthan gum, guar gum, propylene, sodium citrate, and datem of mono- and di-glycerides. These are all red flags on our path to optimal cognitive fitness.

GLUTEN: IS IT REALLY AS BAD AS WE FEAR?

Last but not least, let's talk about gluten. Gluten is a protein contained in a variety of grains including wheat, rye, and barley, which has made the headlines lately with regard to having potentially harmful effects on brain health.

There are many things we don't yet know about gluten consumption and its effects on human health. What we do know is that gluten can have a negative effect on *gut* health. Some people have particularly strong reactions to it, especially those patients with celiac disease who are genetically susceptible to gluten and must avoid foods that contain it. In these patients, gluten causes increased intestinal permeability, creating the leaky-gut environment described earlier, with subsequent symptoms of a weakened immune system and inflammation. Similar reactions are sometimes observed in people who don't have celiac disease, possibly due to negative interactions with their microbiome. In the end, the way your gut reacts to gluten depends on your DNA, both human and microbial, so you need to listen to your body and respond accordingly.

It is much less clear whether gluten is in any way connected with *brain* health. I'm often asked if gluten is bad for your brain and if it should be avoided. Currently, there is no conclusive evidence of a connection between gluten consumption and cognitive decline or dementia. To check on this yourself, you can use the same tool scientists use to look for peer-reviewed publications: PubMed (www.ncbi.nlm.nih.gov/pubmed). If you search "gluten and Alzheimer's or dementia," you will find the most current information on the subject. Be careful to include only journals with English titles, such as the *Journal of Alzheimer's Disease* or *Neurology*. As of March 2017, there are only ten papers or so that look into the relationship between gluten and cognitive disorders, mostly with regard to patients with celiac disease. To give you a sense of what scientists consider a much more significant finding, run a search on "glucose and Alzheimer's or dementia" instead. With this pairing, you'll find close to four thousand papers that report significant and beneficial associations. Given the relationship between the gut and the brain, further clues might come up in the future as more scientists start looking into gluten as a possible risk factor for cognitive impairment. For the moment, the jury is still out. What I can tell you is that so far there is no evidence that eating grains will make you forget names or lose your keys.

While gluten has not been proven to harm our brains, the absence of fiber has. There is plenty of evidence that fiber deficiencies negatively affect the microbiome and therefore, to some extent, the brain. As previously discussed, we also need fiber to stabilize our blood sugar levels and to support a healthy immune system. Since gluten is found in the many grains and cereals that contain fiber, eliminating gluten from the diet stands to compromise an adequate fiber intake. As such, I recommend caution before eliminating grains from your diet. Here's my commonsense approach: find out what works for you. If you are among those who tolerate gluten, carefully chosen, organic, non-GMO whole-grain foods are an important addition to a gut-healthy, and subsequently brain-healthy, diet.

But if you're concerned, speak to your doctor about being tested for gluten allergies or sensitivities. Should you turn out to have a negative reaction to gluten, by all means watch your gluten intake. It's important to assess our choices wisely. In this case, naturally gluten-free grains include amaranth, buckwheat, millet, rice, sorghum, teff, and quinoa (technically, quinoa is a seed). Tofu is also naturally gluten-free—but your average soy sauce is *not*. What most people don't realize is that gluten is present in many foods and products besides grains. Take a look at Table 8 for a list of the many unexpected yet common places we find gluten.

As we reach the end of this chapter, the bottom line is: every person's diet should reflect their genetic uniqueness with the goal of optimizing their health and mitigating risk of disease. Although this field is just developing, we already have the necessary tools in place to assess the effects of various foods on our DNA, human and non-human, and how their interactions support or affect our brains. Also keep in mind that nutrients alone won't achieve our aims. So many of our anxieties around diet take the form of our search for the *perfect* nutrient, the ultimate superfood, or the one magic pill that will cure all our ills. We tend to obsess over the specific qualities of various foods and supplements, mull endlessly over proteins versus carbs versus fat, and debate which fish oil supplement to take. But as we've learned, no matter what we seek, the answer ultimately lies in the nutrients existing in unadulterated versions of the foods we eat. And they can only reach us when we actually pick up the food and eat it. How we eat and how we approach food is really what matters. In order to change our diets, we can begin by relearning the art of eating, which is a question of lifestyle as much as it is nutrition.

Our next step is to take a closer look at those who have indeed mastered the art of eating right for their brains as well as for their overall well-being. Let's turn now to those that are the exception to the rule: the centenarians.

Food Item	**Examples**
Grain cereals	Wheat, rye, barley, oats (unless gluten-free)
Wheat derivatives	Wheat berries, durum, semolina, farina, farro, graham
Malt and malt derivatives	Malted barley flour, malted milk or milkshakes, malt extract, malt syrup, malt flavoring, malt vinegar
Pasta	Wheat pasta, raviolis, dumplings, couscous, gnocchi, noodles (except rice noodles)
Baked goods	Bread, pastries, crackers, cookies, croutons, pizza
Breakfast foods (often contain malt extract/flavoring)	Breakfast cereal (including corn flakes and rice puffs), granola and granola bars made with regular oats, pancakes, waffles, French toast, crepes, biscuits, granola bars

(continued)

Food Item	Examples
Sauces, gravies (often contain wheat flour as a thickener)	Soy sauce, cream sauces made with a roux
Processed meats	Deli meat, cold cuts, pastrami, salami, bologna
Condiments	Salad dressings, marinades, mayonnaise, ketchup
Fried foods (often made with batter containing wheat flour)	French fries, fried chicken, chicken nuggets, fast foods, doughnuts, fried baked goods
Candy and candy bars	
Creamers and nondairy creamers	
Soup, commercial bouillon and broths	
Meat substitutes made with seitan (wheat gluten)	Vegetarian burgers, vegetarian sausage, imitation bacon, imitation seafood
Eggs served at restaurants (might contain pancake batter)	Omelets, scrambled eggs, frittata
Beverages	Beers, ales, lagers, malt beverages that are made from gluten-containing grains, wine coolers, vodka (unless gluten-free)
Drug fillers (often contain starch)	Some drugs, over-the-counter medications and dietary supplements
Lipstick, lip gloss, and lip balm (often contain starch)	
Any foods cooked in shared food-preparation equipment (pasta pot, toaster, deep fryer)	

Table 8. Common gluten-containing sources and where to find them.

9

The World's Best Brain Diets

THE BLUE ZONES

Sometimes it is worth leaving the research lab to find out what works in the real world. By doing so, researchers have discovered entire communities of centenarians—those who are one hundred years old and counting. What's even more interesting is that they have somehow remained as sharp as a tack.

As of today, five regions have been identified that possess the highest concentration of centenarians in the world. These are known as the "blue zones." The first of these longevity hotspots was discovered in the provinces of Nuoro and Ogliastra in Sardinia (Italy), which is the location with the highest concentration of male centenarians in all the world. This is quite a pedigree. In fact, women typically live longer than men, and male centenarians are especially rare. Next up is the Greek island of Ikaria in the Aegean Sea, cleverly nicknamed "The

Island Where People Forget to Die." The third blue zone is Okinawa, sometimes referred to as "Japan's Hawaii," home to the world's longest-living women and to as much as 15 percent of the world's supercentenarians (110-plus years). Crossing the planet, we find the Nicoya Peninsula in Costa Rica—home to 100,000 mestizos with a lower-than-normal rate of middle-age mortality. Finally, the Seventh-day Adventists community in Loma Linda, California, boasts a life expectancy that exceeds the American average by a decade.

In all the blue zones identified so far, people reach age one hundred at rates ten times greater than the U.S. average. And not only do they live longer, they also enjoy remarkably full lives with a very low incidence of heart disease, obesity, cancer, and diabetes—not to mention dementia. Clearly they're doing something right.

Despite the wide geographical distance that separates these regions, or the fact that these cultures couldn't be more different from one another in so many ways, people who hail from these blue zones turn out to lead surprisingly similar lifestyles.

First, they move a lot. In spite of their old age, they incorporate physical activity naturally into their daily lives, like gardening or walking, but also more rigorous activities such as farming, manual harvesting, and even shepherding livestock. They have low stress levels and keep a slow pace of life. In spite of this, they still insist on taking time out to relax, for example, by taking regular naps. Blue zoners also tend to have strong family and social connections, and often belong to religious communities that further reinforce these behaviors. In addition, they have a strong sense of life's purpose and belonging that keeps them socially active and well integrated in their communities. This couldn't be more different from the United States, where frequently elderly parents retire to a different state than their families or to senior homes, sometimes located at a distance from their family members. In the blue zones, grandparents play an essential role in the upbringing, education, and caring of grandchildren, and are often actively engaged in civic volunteering as well. Incidentally, the word *retirement* doesn't exist in the traditional Okinawan dialect.

As for what they're eating, it turns out that blue zoners, no matter their far-flung locations, tend to follow similar diets. While regional variations exist, their typical diets are largely based on plant sources, and are characterized by moderate caloric intake and small portions. Confucian ideals like eating only enough food to feel 80 percent full are at home within these communities. Typically, these centenarians start their day with a large breakfast, followed by a good lunch and a small, often early dinner to facilitate sleep. From a nutritional perspective, they consume a high-carb diet with moderate-to-low levels of protein and fat. Legumes such as beans are a staple dish. Meat is consumed rarely, on average five times a month and in very small portions by our Western standards. Alcohol intake is modest, no more than one to two glasses a day, most often of wine.

For a closer look at their typical meals and recipes, we need to look at each blue zone individually. In Sardinia and Ikaria, people enjoy a Mediterranean diet plentiful with wild, bitter greens like dandelions and grape leaves, legumes like garbanzo beans, and potatoes. They also enjoy and eat plenty of fish—simply grilled and seasoned with herbs like thyme, dill, sage, and marjoram—and the occasional bite of cheese such as Feta and Pecorino. Their beloved olive oil features prominently in their day-to-day lives.

The diet of the Okinawans couldn't be more different from that of the Mediterranean countries, yet it turns out to be just as delectable. Some classic staple foods include their bright purple sweet potato, various seaweeds, vegetables and fruit like bitter melon—along with the use of soy products like tofu and natto (fermented soy beans). Of course, fresh-caught fish is also a main staple, as are brown rice, green tea, shiitake mushrooms, ginger, and garlic. Among these Okinawans, almost no meat, eggs, or dairy products are consumed. Additionally, this diet is particularly low in calories, even by Japanese standards. A typical Okinawan centenarian consumes 20 percent fewer calories than the average Japanese citizen, highlighting caloric restriction as a proponent of increased life spans.

An entirely different ingredients list hails from the Nicoya Peninsula in Costa Rica, where centenarians regularly enjoy the three main staples of Mesoamerican agriculture: beans, corn, and squash. Homemade corn tortillas accompany most meals. Black beans, white rice, yams, and eggs are dietary staples that the culture combines with a variety of fruits like mango, passion fruit, guava, papaya, and their distinctive peach palms high in vitamins A and C. These blue zoners also consume fish, as well as some meat. Finally, Costa Rica is known for its excellent coffee, and the people of Nicoya are the third blue zone (along with Sardinia and Ikaria) to drink it daily.

Last but not least, Loma Linda in California hosts a large community of Seventh-day Adventists, a Protestant Christian denomination that encourages members to eat a well-balanced vegetarian diet with plenty of legumes, whole grains, nuts, fruits, and vegetables. Their top foods include avocados, nuts, beans, oatmeal, whole-wheat bread, and soy milk. Some Seventh-day Adventists eat eggs and dairy. When it comes to drinking, their only beverage is water. No coffee, tea, soda, or caffeinated beverages are permitted. Sugar is a no-no, too, except via natural sources such as honey. It comes as little or no surprise that these blue zoners, besides living longer, also have the lowest rates of heart disease and diabetes in the United States, as well as a very low rate of obesity.

THE MEDITERRANEAN DIET

For many, one of the first diets that comes to mind when talking about brain health could very well be the Mediterranean diet. Centenarians in two blue zones out of a total of five eat this way. Researchers have long praised the Mediterranean diet for promoting brain health as well as overall physical health. In fact, as famously heart-healthy as this diet is, it also benefits your brain. A large body of scientific literature, my own work included, shows that people who closely follow a Mediterranean diet are not only less likely to develop diseases like diabetes, obesity, and cardiovascular disease, but also have a reduced risk of cognitive impairment and Alzheimer's as they age.

Born and raised in Italy, I have firsthand experience of what this diet is all about. For Italians, it is not even a diet so much as it is a way of eating and experiencing food. If you were to travel from Italy to the Greek Islands, or from the southwest of France to Barcelona, you'd probably notice a wonderful variation of cuisines—a wide spectrum of specialties, different key ingredients along with all the local habits that accompany them—each and every one a source of regional pride. However, what all these countries have in common is a healthy respect for fresh, locally grown, sun-basked products.

If you were to create a Mediterranean diet pyramid, as many nutritionists are fond of doing, you'd find a wide range of vegetables, fruits, beans, and nuts at the base, as they are the main focus of the plate. Whole grains like wheat, oats, spelt, and barley (consumed in minimally processed forms to provide the maximum dose of nutrients), often served along with wild-caught fish, from trout to *orata* (dorade, sea bream), are one step up and are eaten fairly often. Meat and dairy are occasional indulgences. Herbs and spices are used freely to naturally flavor the food, reducing the use of extra fat and salt in the process. Sweets are consumed in small portions, usually as a Sunday treat or at special celebrations. In addition, they tend to be much healthier than the usual supermarket-bought desserts, since they are typically made using nuts and seeds and sweetened with honey, molasses, and other natural sugars. As a whole, the Mediterranean diet is a very fresh, very tasty diet, low in calories and fat, and rich in all sorts of brain-essential nutrients.

Olive oil deserves special mention. It is now believed that regular consumption of extra-virgin olive oil is a prime reason for the health effects of the Mediterranean diet. Olive oil that is truly extra virgin has a distinctively bitter, almost pungent taste, and owes its reputation as "the world's healthiest oil" to its high antioxidant content. In fact, this oil contains heart-healthy monounsaturated fat blended with artery-scrubbing *phenolic* compounds and vitamin E, another important antioxidant. This particular combination makes olive oil practically magical, as polyphenols also protect and preserve the delicate vitamin E. Even clinical trials show that if we regularly consume extra-virgin olive oil (up to 1 liter/1 quart a week), we can actively protect ourselves from cognitive decline.

Red wine is another main staple of the Mediterranean diet and an excellent source of anti-aging antioxidants. Mediterranean people have a curiously relaxed relationship with wine. Wine is even given in sips to children to introduce them to the fine art of drinking it. I remember my dad giving me my very first sip of red wine diluted with water when I was only six years old. Before you balk, I should let you know that to this day I have never suffered from a hangover. This is because we learn not only how to handle wine at a young age, but also how to consume it within specific guidelines. For men, up to two small glasses of wine a day is considered ideal. Since women absorb alcohol more rapidly than men, one glass a day does the trick. It's important to note that wine delivers this benefit when drunk as an accompaniment to the meal rather than as a drink had by itself. It is considered downright detrimental to drink wine on an empty stomach, for example.

Another special feature of the Mediterranean diet is its social component. One doesn't eat (or drink) alone, nor while walking to work or in a mall, let alone while sitting in front of a computer screen. Rather, meals are consumed in the company of others and savored over enjoyable conversation—which, funny enough, often revolves around food. My grandmother would start planning dinner while we were having lunch, and Sunday's lunch menu was often discussed as early as at Tuesday night's dinner. This invited everyone to put in their two cents, or in

Italian, "*dire la loro*," while at the same time making people more aware of their eating choices. "How about *pasta al pomodoro*? Wait, we had it last week—how about polenta instead?"

Finally, daily physical activity is second nature to Mediterranean cultures. This doesn't mean strenuous exercise, however. Traditionally, the average person does not go to a gym to work out. Rather, leisurely activities such as walking, housework, gardening, riding a bike, or taking the stairs instead of the elevator are all normal parts of the culture's daily routine. Truth be told, the majority of buildings in these countries don't even have elevators.

All in all, the Mediterranean diet is not so much a diet as it is a lifestyle. It is vibrant and fresh, and genuine foods combined with regular exercise, a rich social life, and a positive outlook all contribute toward the long life span of the Mediterranean people.

What is really exciting is that the health benefits of the Mediterranean diet even show up on brain scans. Do you remember the MRI scans in chapter 1? Those scans come from a series of brain imaging studies in which we looked at the effects of the Mediterranean diet in over fifty participants ages twenty-five to seventy-plus years old. The results were astonishing. Regardless of their age, those who followed the diet had overall healthier brains than those on a typical Western diet (or any diet full of red and processed meats, sugary beverages, and sweets and low in plant-based foods and fish). The brains of those on these less healthy diets seem to literally age and shrink more rapidly. In some studies, their brains appear to be a good five years older.

Not only were these beleaguered brains shrinking. Their activity was also reduced. Worse still, even though none of the participants had yet to demonstrate any outward sign of cognitive impairment, the brains of those on a Western diet were already carrying more amyloid plaques than was normal for their age, indicating a higher risk of developing Alzheimer's in the future.

The good news is that while you probably get the biggest payoff adopting such a diet early on in life, research shows that it is never too late to reap the benefits of a healthy shift toward better lifestyle choices. For example, a study of over ten thousand women showed that those who followed the Mediterranean diet during middle age, though not necessarily prior to that, were much more likely to live past the age of seventy without chronic illness or mental deficiencies than those who did not eat as healthily.

Luckily, you don't have to move to these Mediterranean countries to keep your mind sharp. A new diet known as the MIND diet (Mediterranean-DASH Intervention for Neurodegenerative Delay diet) makes the Mediterranean diet easier to follow no matter where one lives in the world. Its core principles are: three servings of whole grains plus a salad and one additional vegetable every single day, along with a glass of wine. Legumes are eaten every other day, while poultry and berries are included twice a week. Fish is consumed once a week. Additionally, to have a real advantage against the devastating effects of Alzheimer's, dieters are encouraged to limit foods considered unhealthy, especially fried or fast foods but also high-fat dairy and meat. If that sounds too demanding, here's some incentive: the MIND diet lowered the risk of Alzheimer's by as much as 53 percent in participants who adhered to the diet rigorously. But even those with "middle-of-the-road" compliance with the diet still showed a 35 percent risk reduction. Finally, if the Mediterranean diet isn't your thing, perhaps Chinese food is?

NOODLES OF LONGEVITY

Even though this next region has yet to be granted blue zone status, Bama Yao in southern Guangxi, once one of China's poorest regions, is home to the famed Longevity Village, where many people live to one hundred and beyond.

Geographically speaking, Bama is surrounded by picturesque hills and mountains, with the Shangri-la's Panyang River flowing in between. Thanks to its clean, fresh air, Bama is hailed by many as a natural oxygen tank. It is in this idyllic location that Bama centenarians lead a lifestyle worthy of any other blue zoner. They eat frugally and mindfully, favoring freshly picked vegetables and fruits above all other food. Vegetables in particular are part of each meal—breakfast, lunch, and dinner. Other staple foods include rice and hominy (maize kernels) with the addition of sweet potatoes, fruits, nuts, and seeds. Sweet corn, beans, peas, lentils, and fresh-caught fish round out their diet. Hemp seed oil, a vegetable oil very high in PUFAs, is another main staple. Overall, these centenarians follow a low-calorie, low-fat diet, which is high instead in carbohydrates, vitamins, minerals, and fiber.

In addition, they are mostly farmers and engage in this work no matter their age. Traditionally, this remote area had no access to mechanical equipment, power tools, or even electricity until very recently, so nearly everything was done entirely by hand. And never mind watching TV or spending hours online attending to social media. This society thrives on a "real-time" social network, proving once again that such exceptional well-being anchors itself in a strong sense of community and the sense of belonging that it provides. In addition, elders are held in the highest regard. One sign of this is that families serve their elders first at every meal. Further, everybody turns to their grandparents for sage advice.

Speaking of China, several herbs traditionally used in Chinese medicine deserve special attention, as they are some of the world's most renowned brain tonics. One of the oldest plants on the planet, ginkgo biloba, has long been known for its potential to treat age-related mental decline, so much so that it's often prescribed in countries like Germany and France. It is believed that ginkgo works by thinning the blood and thereby improving oxygen flow to the brain. Although the results are not unanimous, some clinical trials showed that administering 240 mg/day of ginkgo extract for about six months had beneficial effects on attention, memory, and overall cognitive function.

Ginseng is another herb with celebrated anti-aging properties to the point that it is considered the Fountain of Youth among the Chinese. Though more data is needed, a few clinical trials demonstrated that supplementation of 4.5 grams a day of Panax ginseng might be helpful in improving cognitive function even in Alzheimer's patients.

INDIAN CURRIES

India has a spectacularly low incidence of Alzheimer's as compared to more developed countries, even after accounting for their lower life expectancy rates. As a comparison, Americans are eight times more likely to get Alzheimer's than their Indian counterparts.

Research indicates that diet has a lot to do with this. In fact, Indian cuisine is particularly rich in spices known for their brain-protective properties. It turns out that turmeric, the signature spice of Indian cuisine, is a powerful antioxidant and anti-inflammatory agent. A mustard-yellow powder eaten daily by Indians in their curries, turmeric has been used for at least five thousand years in Ayurvedic medicine against many types of pain and inflammation associated with aging. Recent evidence shows that this spice, or more specifically, its active ingredient *curcumin*, helps protect against cognitive loss and dementia by keeping our neurons healthy as we age.

For example, in several lab studies, mice that were fed curcumin developed fewer of the amyloid plaques associated with Alzheimer's than animals that weren't. Further, older animals who already had plaques in their brains experienced a significant reduction in the number and severity of these plaques. In other words, curry seems to help the brain stay clear of Alzheimer's lesions.

To date, only a few clinical trials of curcumin supplementation have been completed on humans, yielding negative or inconclusive results. However, as a large number of investigators believe in its anti-aging potential, there are several ongoing trials evaluating curcumin's efficacy against aging and dementia. It is possible once again that eating the actual spice might prove more synergistically powerful than taking an isolated ingredient.

THE ANTIOXIDANT DIET

As we have seen in the previous chapters, as the brain ages, it employs antioxidants to fight off harmful free radicals. The antioxidant diet aims at increasing the intake of foods and nutrients with high antioxidant potential based on the idea that the more antioxidants you have at your disposal to help squelch free radicals, the lower your brain's risk of suffering from oxidative stress and disease. This diet can be seen as a spin-off of the Mediterranean diet that puts even more emphasis on the nutritional content of plant-based foods.

The plant kingdom is abundant in especially powerful antioxidants such as vitamins C, E, and beta-carotene; the mineral selenium; and several phytonutrients—like the *carotenoids* found in orange vegetables (carrots, sweet potatoes) and the *anthocyanins* that give cherries their bright red color. Berries like blackberries and blueberries, citrus fruits such as lemons and oranges, along with Brazil nuts, walnuts, and many dark-colored beans like raw cacao, are loaded with natural antioxidants that can help protect the brain from harm. Vegetables, in particular spinach, peppers, and asparagus, are also excellent sources of antioxidants, as are some oils like extra-virgin olive oil. These are not ordinary foods. These are superfoods—which should be routinely added to our diets, no matter our age.

And now for the trick up the neuro-nutritionist's sleeve: *glutathione*. Glutathione is known as the "master antioxidant." In a way, it is the supervisor of all other antioxidants, and is also in charge of your body's detoxification and immune system. As such, it is *the* antioxidant that everybody needs to stay healthy and prevent disease. Yet many people have never heard of it. Glutathione is produced internally in the body, but there are several foods and supplements that help to boost its levels. Foods rich in sulfur, especially onions, garlic, asparagus, avocado, spinach, and cruciferous vegetables such as broccoli, cabbage, and cauliflower, are all very helpful in raising your glutathione levels.

In addition to increasing our dietary consumption of antioxidants, it is important that we stay away from those foods that further deplete our brains' antioxidant potential. In the past decade, scientists have shown that some foods contain high amounts of *advanced glycation end-products* (AGEs), harmful compounds that, much like free radicals, can produce inflammation and negatively affect nearly every type of cell and molecule in the body. This in turn accelerates brain aging, cognitive decline, and disease.

Animal-derived foods high in fat and protein, such as butter, margarine, sausage, hamburger meat, and pork chops, contain a lot of AGEs. Additionally, they are prone to additional "AGEing" during cooking, especially with dry heat. Good examples of this would be broiled beef frankfurters and fried bacon, which are quintessentially harmful AGE-carrying offenders. Recommended cooking methods to lower your chances of oxidative stress are (1) steaming, (2) using shorter cooking times, and (3) cooking at lower temperatures. So if you are in the mood for protein, poached eggs and steamed salmon are nutritious low-AGE choices. Also, cooking with acidic ingredients such as lemon juice or vinegar actually manages to reduce the AGE levels in animal foods. Have you tried roasting your chicken in balsamic vinegar? It's quite delicious.

In contrast to animal foods, carbohydrate-rich foods contain relatively few AGEs, even after cooking. Vegetables like carrots and tomatoes, fruits like apples and bananas, and whole grains such as oats and rice are at the top of the AGE-free list. With a healthy supply of free-radical-neutralizing foods in your diet, your brain will be armed to both withstand and ward off the age-related "rusting" effects of oxidation and fight off disease.

CALORIC RESTRICTION AND THE KETO DIET

Although not as widely publicized as the Mediterranean diet and certainly not as enticing, caloric restriction, or dramatically reducing your calories within reason, has been associated with increased longevity and improved cognitive function.

The strategy behind this diet is based on almost a century of scientific data showing that stressing our bodies via caloric restriction pushes our cells to get stronger and more resilient against stress. "That which does not kill us makes us stronger," as put by Nietzsche quite elegantly in his philosophical essays. Just as muscles get stronger the more you exercise, your brain cells strengthen themselves when they are hungry.

For reasons under investigation, caloric restriction boosts the brain's antioxidant defense system in laboratory animals. Plus, it ramps up the action of mitochondria (the energy factories of the cell), producing more energy. It also reduces inflammation, prevents deposition of Alzheimer's plaques, and seems to promote *neurogenesis*—the formation of new memory-related neurons. That's quite an impressive résumé.

In general, these effects are observed in animals whose diet has been limited to 30 to 40 percent of their habitual caloric intake. To compare, when eating a calorie-restricted diet, instead of consuming an average of 2,000 calories a day, one would bring that number down to 1,200 to 1,400 calories a day. While limited data is available in humans, a recent clinical trial showed that a similar caloric restriction does indeed lower the risk of memory loss. This study looked at fifty healthy normal-to-overweight elderly subjects, a third of whom were placed on a calorie-restricted diet. After three months of intervention, their memory performance had improved by 20 percent. Those who stuck to the diet more closely also showed markedly improved insulin levels and reduced inflammation.

However, it turns out that although reducing the overall amount of calories definitely helps, *fasting* might work even better. If you just thought to yourself "forget it," picturing a guru perched upon a bed of nails, don't worry—we're not talking about prolonged fasting but *intermittent* fasting. Intermittent fasting is a type of limited, shorter-term fasting that involves short spells of dietary restriction amid longer periods of habitual eating. This seems to confer the most health benefits. For instance, intermittent fasting can increase the life span of laboratory animals by up to 30 percent. This makes sense if you consider that most animals, including humans, evolved through multiple short-term periods of caloric restriction, for example, during winter. As a result, our metabolism operates more efficiently when freed from the burdens of 24/7 digestion and nutrition assimilation.

While more work is needed to prove the potentially beneficial effects on cognitive health due to fasting, there is evidence that a form of intermittent fasting known as the "5:2 diet" has positive effects on cardiovascular function, and therefore might help slow age-related cognitive decline. This diet involves eating normally five days a week and then taking in a maximum of approximately 600 calories a day for the following two days. In a recent study, 107 overweight or obese women were divided randomly into two intervention groups. One group was put on a calorie-restricted diet (1500 kcal/day for seven days) and the other group on the 5:2 diet. After six months of dieting, both groups had lost weight and showed reduced inflammation, insulin resistance, cholesterol, triglycerides, and blood pressure levels. However, these improvements were more pronounced in the 5:2 group than in those on the stricter seven-day version, indicating that reduced-calorie dieting for just two days a week might be just as good if not better than 24/7.

Another bonus of calorie restriction is that fasting increases production of ketone bodies. As mentioned before, ketones are the only alternative energy source for the brain when our glucose supply runs too low. Since many people might find fasting difficult, supplementation of ketone bodies via a low-calorie diet has been proposed as a viable alternative to support brain health.

The high-fat, low-carbohydrate "ketogenic diet" was formulated in the 1920s and is based on the principle that if one drastically restricts carbohydrate intake, the body goes into a state of *ketosis*, forcing it to burn fat, which in turn produces ketones. Besides being helpful for weight loss, this diet is known for its anticonvulsive properties and is widely used to treat epileptic seizures, revealing a brain-protective effect.

Recent data suggests that the keto diet might also help with diseases such as Parkinson's and Alzheimer's. Although clinical trials have been scarce, preliminary studies have shown that supplementation with medium-chain triglycerides (MCT), a type of fat that is among the best sources of ketones, improved Parkinson's symptoms by 43 percent after just one month of treatment. Similarly, patients with Alzheimer's or MCI showed improved cognitive performance after a few months of supplementation with caprylidene (Axona), a prescription medical food that the body metabolizes into ketone bodies. These studies are limited however by a very small sample size and await replication. It also remains unknown whether eating natural foods rich in MCT, like coconut oil, could be just as effective.

If the keto diet sounds appealing to you, it is important to keep two things in mind. First of all, ketone bodies are *not* the preferred energy source for the brain. As we discussed, the brain always needs, at a bare minimum, 30 percent of calories from glucose to work efficiently. Second, this diet is basically the opposite of the scientifically proven Mediterranean diet. Third, increased fat consumption can alter the body's metabolism. Lastly, even though your body will eventually burn off the saturated fat ingested as part of this diet, your cholesterol level might increase in the meantime. Plus, fat-rich foods are usually low in fiber, which is hard on your digestive system, and rich in protein, which might be hard on your kidneys. As a result, adverse effects such as constipation, flatulence, *dyspepsia* (disturbed digestion), and "keto breath" (bad breath) can be fairly common.

LESSONS LEARNED FROM THE WORLD'S HEALTHIEST DIETS

What lessons can we learn from the world's healthiest diets? How can we incorporate their principles into our everyday lives when we're constantly tempted by processed foods and excess sweets during the long hours often spent at our desks, stressed and restless for something more?

While all these diets might at first glance have very little in common (seaweed in Okinawa, olives in Sardinia, curry in India)—they actually share a key common ingredient. With the exception of the keto diet, each of them provides an excellent example of a whole, nutrient-dense diet known to benefit the brain as much as the rest of the body.

In each of these diets, regular consumption of wild, fresh greens is integral. These greens come with an arsenal of vitamins, minerals, and antioxidants that brain cells need to stay healthy and communicative. Fresh fruit, picked ripe from the trees, is another excellent source of vitamins as well as natural sweetness that at the same time curbs cravings for refined sugars. Among all fruits, berries seem to be especially brain-supportive. Many research studies have shown that berry extracts from blueberry, cranberry, blackberry, cherry, strawberry, and Concorde grape all ameliorate or even prevent cognitive declines in laboratory animals.

Although many of us love chocolate, few of us realize that raw cacao *also* comes from berries. Cacao is loaded with antioxidants like *theobromine*, a close relative of caffeine, and many powerful flavonoids. A recent clinical trial showed that consumption of cocoa drinks with a high flavonoid content of 500 to 1000 mg improved attention and memory in the elderly while reducing inflammation and insulin levels in as little as eight weeks.

And what about coffee? Coffee comes from roasted coffee beans, which are again the berries of the *Coffea* plant. As most people are aware, coffee beans contain caffeine, a substance that keeps you awake at night but in addition possesses fierce antioxidants such as *chlorogenic acid*. It is worth noting that even though coffee and cocoa drinks are not consumed in all blue zone communities, those who do drink them regularly have even lower rates of diabetes and heart disease. While results are not always consistent, some research studies have shown that people who drink coffee daily in midlife are less likely to develop dementia when they get older. Again, everything in moderation. Too much coffee might affect your heart rate as well as your sleep quality.

For those of us who love wine, we celebrate that grapes are berries, too. Red wine is a great source of *resveratrol*, an aromatic compound found in the skin of grapes (but also in raspberries and mulberries), which is well-known for its antioxidant and neurons-protective properties. Wine also contains flavonoids that protect blood vessels and heart health. While pretty much everybody agrees that one to two glasses of red wine a day are a key part of aging gracefully, clinical trials have so far failed to show the beneficial effects of resveratrol on cognition. Again, this raises the question as to whether taking in these benefits via our food (or better, our wine) is in fact more powerful than attempting to glean results via supplementation.

While not all centenarian communities drink tea, there is some evidence that this popular beverage might also help protect brain cells and fend off dementia as we age. Most people who consume tea regularly choose black tea as their favorite tea. However, the brain prefers green tea. Green tea contains twice the amount of antioxidants than black tea and is therefore a more powerful anti-aging ally. Green tea is also quite rich in a special flavonoid called *EGCG* (*epigallocatechin-3-gallate*) that appears to protect the brain from accumulation of Alzheimer's plaques.

Nuts and seeds are another staple food of many centenarians. These pint-sized nutritional dynamos are loaded with healthy unsaturated fats, protein, fiber, and a variety of antioxidants. Walnuts in particular are well-known for their high level of PUFAs and antioxidants like vitamin E, melatonin, and *ellagic acid.* These nutrients act synergistically to enhance the effects of the PUFAs, all the while increasing absorption of their own protective compounds. The result is improved cognitive function, at least when tested in aged animals.

Local whole grains, beans, and starches are also dietary staples of most longevity diets. These foods provide a slow release of brain-supportive carbs and fiber while reducing the meal's glycemic load, avoiding sugar rushes and crashes. Sweet potatoes in particular are part of most longevity diets. Not only are they full of dopamine-enhancing nutrients, but they also contain high amounts of one of our brains' very favorite antioxidants, beta-carotene, which we convert into vitamin A. A sweet potato alone provides 368 percent of the recommended daily dose of vitamin A, which our bodies can store away for times of need.

Unprocessed, high-quality vegetable oil and fish rich in unsaturated fat are also common in most longevity hubs. The nutrients contained in these foods help promote cholesterol transport, which protects the heart while ensuring a healthy supply of oxygen and nutrients to the brain. Additionally, fatty fish like salmon is among the best natural source of brain-essential DHA. To date, as many as nine large-scale epidemiological studies have concluded that regular fish consumption is crucial for brain health. Most studies reported that middle-aged and older people who consumed fish regularly succeeded in delaying cognitive decline and reduced their risk of Alzheimer's by up to 70 percent as compared to those who ate little to no fish. They granted themselves this insurance policy by eating high-quality fatty fish just once or twice a week.

Another important lesson is that it's not only about what you *do* eat, but also about what you *don't.* With the exception of the keto diet, all longevity diets are characterized by infrequent consumption of red meat and dairy, thereby lowering the intake of saturated fat and cholesterol. This expedient alone might very well account for the centenarians' lower risk of heart disease. When they *do* eat meat and dairy products, these come from pasture-raised animals (oftentimes goats and sheep), whose meat is leaner and higher in PUFAs than that of domesticated animals, and whose milk contains higher amounts of brain-essential nutrients like B vitamins and serotonin-boosting tryptophan.

We also saw that dessert is considered an exceptional treat and never the norm. In addition, the use of natural sweeteners like raw local honey, molasses, and dried fruit renders refined-sugar products altogether unnecessary for these populations. Most centenarians as well as their younger family counterparts do not drink or even like soda, one of modern society's most hidden sources of added, excessive sugar. To this day, I have never seen an Italian *nonna* (grandmother) drinking Coca-Cola—unless she's up to some mischief!

Overall, tradition and science agree that there are common dietary principles that promote longevity, deeply rooted in the choices we make as well as our lifestyle habits as a whole.

10

It's Not All About Food

FIT, ACTIVE, AND HEALTHY

Keeping the brain stimulated, whether physically or mentally, is a life-long enterprise that can continually increase its cognitive reserve, allowing the brain a greater flexibility to tolerate age-related changes without developing memory loss and other cognitive difficulties. Participation in sports and leisure activities, higher education, intellectual exchange, work complexity, socializing with friends and family, and even our sleep—all contribute to our ability to sustain cognitive function well into our old age, giving us sharp memories and reducing the risk of Alzheimer's.

With the minimum of drawbacks and plenty of benefits, a well-rounded, healthy lifestyle can improve our general health, protecting and supporting our brains over the course of a lifetime. In this chapter, we will explore which specific physical exercises, intellectual and social activities, and even sleep habits are necessary to keep our brains functioning at peak performance level.

BE SMART: EXERCISE YOUR HEART

The rumba and the cha-cha-cha . . . horseback riding and even snorkeling. While these might not be the first things that come to mind when contemplating how to keep your brain in top shape, they might very well be the ideal prescription.

Exercise has been touted to be a cure for nearly everything, from menstrual cramps and osteoporosis to obesity, type 2 diabetes, heart disease, and depression. It is also the latest addition to the growing list of lifestyle factors that help protect our brains against disease.

However, the evidence that exercise provides substantial benefits for the brain has yet to be fully accepted by the mainstream medical community. For example, if you were to see a neurologist with concerns about memory loss, it's unlikely you'd walk out with a prescription for physical therapy or exercise. Even the most enlightened of doctors would be hard-pressed to recommend a specific fitness regime as an answer to your prayers. Should I run every day? Lift weights? Take a Pilates class? The truth is, there are still no uniformly established medical recommendations for "brain fitness."

But we are getting there. An emerging body of scientific literature is documenting the beneficial influence of physical activity on the brain as well as on the body. The physically fit elderly typically perform better on reasoning and working memory tasks, and their reaction time is also quicker than that of the sedentary elderly.

There are many good reasons for your brain to enjoy exercise. First, exercise promotes heart health—and as we discussed before, what's good for the heart is good for the brain. Physical activity, especially *aerobic* (the kind of exercise that makes your heart beat faster), enhances blood flow and circulation, improves the delivery of oxygen and nutrients to your brain, and also slows down the buildup of plaque in your arteries. This is particularly useful as we age, since our blood flow to the brain would otherwise naturally slow down.

Exercise is also a natural antidepressant. Don't you feel more relaxed and in a much better mood after a workout? Your brain does, too. That's because exercise pumps up your *endorphins*, our bodies' natural painkillers, while increasing production of serotonin, making you feel happier. The famous "joggers' high" is nothing less than exercise impacting the *opioid* system in the brain—the same system that is activated by drugs like opium, a muscle relaxant. Exercise however allows us a natural high as it delivers pain relief, relaxation, and even euphoria, producing an overall sense of well-being.

It doesn't end there. One of the prominent yet underappreciated features of exercise is an improvement in memory performance. Studies have shown that physical activity stimulates memory formation, increases our neurons' ability to recover from injury, and is exceptionally beneficial to the formation of brand-new brain cells. The more you work out, the more your brain produces a protein called *brain-derived neurotrophic factor* (BDNF), which plays a key role in growing memory-forming neurons.

On top of that, physical activity enhances immune system activity, increasing our defenses against disease, and even boosts the enzymatic activity that is particularly effective at dissolving Alzheimer's plaques in the brain, further reducing risk of memory loss and dementia.

To sum it all up, exercising your body does a whole lot of good, not least of all for your brain.

Before we leap into action, let's consider the emerging scientific view of what constitutes brain-boosting exercise in the first place. In general, there is consensus that people who engage in regular physical activity are more likely to remain mentally sharp as compared to those who are sedentary. For example, a study of nearly two thousand elderly participants found that those who engaged in activities such as walking, running, jogging, or bicycling had a 43 percent lower risk of losing their mental capacities as they aged.

However, further studies have shown that as long as you keep active, you might not need to "work out" at all. A number of studies have shown that regular engagement in leisure-time activities (LTA) during midlife can reduce your risk of cognitive decline later in life. Although we don't typically think of these activities as "exercise," whenever you engage in activities that require a certain amount of movement (such as taking the stairs instead of an elevator, going for a stroll in the park, doing house cleaning, or even babysitting), you are working your body as well as your brain. It's not about intensity as much as frequency and consistency. In fact, the study above showed that even those who engaged in *light* physical activity, such as leisurely walking or gardening, had a 35 percent reduced risk of dementia as compared to those leading more sedentary lives, which isn't much less than the 43 percent risk reduction achieved by jogging.

While more strenuous activity might yield greater benefits, many people, especially the elderly or those with injuries, simply can't tolerate exercises like high-intensity training, running, jogging, or spinning. The good news is that working your body *as much as you can*, while keeping active throughout the day in a consistent fashion, is an excellent strategy to boost your memory and age-proof your mind. The goal is to keep moving.

This is crucial, as study after study is showing that leading a sedentary life simply makes your brain age faster. In particular, the memory centers of the brain are known to shrink in late adulthood, leading to impaired memory and reduced mental sharpness. By using brain imaging techniques like an MRI, several teams reported that this shrinkage is much more pronounced in the sedentary elderly than in those who remain active. When my colleagues and I looked into this, we found similar results in people who were in their thirties and forties, indicating that a sedentary life is harmful to your brain regardless of how old you are.

In general, the term "sedentary" refers to people who participate in sports or leisurely activities less than once a week or not at all. If the longest you walk is from the couch to the car, or if you spend more time horizontal (or seated) than vertical, it's time to get up.

I hear some "buts." But what if I've never worked out in my life? But I'm *really* out of shape. But I have bad knees, a bad back, a bad heart! How do I turn all that around?

It's true what they say: it is actually never too late to make a change. Clinical trials show that the mere act of walking can slow down brain shrinkage in just one year's time. That's regardless of whether the participants are used to walking or not. For instance, a study of 120 sedentary adults assigned half of them to a walking program aimed at improving aerobic fitness. The other half was assigned to a toning program that included exercises like yoga or stretching but no aerobic activity. In the exercise program, participants were assigned walking as their sole exercise. They were asked to start by walking ten minutes a day and to do so at a speed slightly faster than their normal pace. Little by little, everybody was able to increase their walking speed and duration until they reached a preset goal of forty minutes of nonstop brisk walking, three times per week. No huffing and puffing were necessary. The pace was that of walking when in a hurry or as if late for a doctor's appointment.

MRI scans demonstrated that this simple exercise regimen had incredible effects on the brain. In older adults, the hippocampus typically shrinks by 1 percent to 2 percent a year, which is what continued to happen in the group that was doing toning rather than walking exercise. But in the exercise group that was walking briskly, the hippocampus *grew* by 2 percent, producing an increase in memory performance.

Therefore, those who did nothing more than walk at a relatively quick pace effectively rolled back their brains' clock by almost two years.

FROM HEART TO HEAD

So far we've seen that whether we're talking about diet or exercise, what's good for the heart is good for the brain. There's a saying in the cardiology community that you're only as old as your arteries. If your arteries age, it wears out your heart, which in turn wears out your brain. And yet, far from just pumping oxygen and nutrients to the brain, the human heart turns out to have more impact on aging than ever imagined. In fact, your heart is secretly helping you remain mentally and physically young in spite of time marching on.

The secret lies within the rejuvenating properties of our blood.

As shocking as this might sound, the rejuvenating properties of our blood have long been recognized, so much so that people have tried drinking blood as an anti-aging treatment for hundreds of years. The idea of refreshing old blood with new harkens back to the fifteenth century, when Pope Innocent VIII allegedly drank the blood of young boys to prevent aging. Legend says Countess Elizabeth Báthory, the most prolific female serial killer in history, murdered hundreds of her young servants so she could take baths in their blood and preserve her youthful looks. Stories of vampires that remain eternally young by feasting on blood have been part of pop culture since the 1700s.

It was only a matter of time before the subject would come under science's scrutiny. In the nineteenth century, scientists started experimenting with a procedure called *parabiosis*—a mismatched joining of dissimilar pairs of animals achieved by stitching together their respective skins. Biology did the rest. Natural wound-healing processes led to new blood vessel growth, sealing the circulatory systems of the animals together and allowing their blood to flow from one to the other.

In the 1950s, a group of scientists in New York City used this method to join the circulatory systems of two mice, one old and one young, which ended up sharing their blood supply. This produced some remarkable results. The blood of the young mouse seemed to bring new life to the aging organs of the old mouse, which grew stronger and healthier. Its heart and lungs started functioning better. Even its coat grew shinier. Less advantageous things happened in the opposite direction, as the young mouse receiving the older animal's blood appeared to age prematurely. In the end, the old mouse ended up living several months longer than average—which is a significant time for a mouse. This suggested that the blood of the younger mouse might very well be responsible for increasing its longevity.

Only just recently this method was used to show that exposure to young blood can indeed perk up the *brains* of older animals. A series of studies showed that when older mice were given blood from their younger counterparts, there was a burst of new neuronal growth in the memory centers of the brain. This in turn improved learning, memory, and endurance in the older animals. Similar results were obtained by injecting blood of young humans into older animals, suggesting that a shot of blood might be the up-and-coming youth elixir of the future.

These discoveries have spurred a quest for a better understanding of what's responsible for this newfound "brain rejuvenation." While we don't yet know for sure how and why these transformations occur, initial studies indicate that *stem cells* might be involved.

What are stem cells exactly?

Stem cells are "mother cells." They're unique in that they are able to develop into any type of cell in the body. Given this ability, they are essential in repairing all sorts of tissues—including brain tissue.

These stem cells are floating continuously in our bloodstream. What scientists have discovered is that, as people get older, their stem cells are still present in blood but are beginning to falter. This is because our blood, in addition to containing precious stem cells, also contains the proteins that are responsible for activating them. These blood proteins—one in particular called *growth differentiation factor 11* (GDF11)—become less efficient as we age, slowing down cell regeneration and possibly contributing to memory decline and neurological deterioration.

These findings offer a potential rejuvenation strategy. It is possible that replenishing these blood proteins with young blood will act as a health booster, consequently improving the birthrate of new cells in the brain. Clinical trials are ongoing to test if blood from young donors can indeed turn back the clock in older people. In the meantime, though, there are some pressing questions that accompany these experiments. Do we really need to go to the trouble of having a blood transfusion in order to keep our memories? Wouldn't it be best to stop our blood from getting old in the first place?

While more research is needed to fully explore the mechanisms behind brain rejuvenation, one thing is clear. The proteins that abound in young blood, which are essential to restoring vigor to an aging brain, are influenced by several factors including, of course, our diet. Several nutrients are thought to enhance the power of these revitalizing proteins. These include flavonoids from fruits and vegetables; antioxidants like vitamins C and E, also found in fruit, vegetables, and seeds; a number of other vitamins, especially vitamin D, which is found in fatty fish, eggs, and milk; and vitamin K, which is abundant in organ meat, fermented soy foods like miso and natto, and also vegetables like dandelion greens. Just a heads up: we'll hear more about dandelion greens in the pages that follow.

That said, keep in mind that healthy blood calls for a healthy heart.

The brain is so intimately dependent on the body's blood for nourishment and sustenance that it is irrigated by no fewer than 100,000 miles of blood vessels. That's the equivalent of six round-trip journeys between New York City and Tokyo. Even though you can't feel it, every single minute the heart pumps 2 pints of blood directly to the head, which is the only way for brain cells to take up all the nutrients and oxygen they need. This brings us right back to where we started. You are only as old as your arteries—and as old as your *brain's* arteries in particular.

I cannot stress enough the importance of keeping your blood vessels as clear and open as possible as a powerful preventative against brain aging and disease. Cardiovascular disease is a major risk factor for dementia, and many people don't realize that it is in large part not only modifiable, but also largely preventable. There are many ways to take care of our hearts, and many have to do with leading a healthy lifestyle.

The prescription is simple. (1) Engage in regular physical activity and you will help your heart stay strong. (2) Eat a diet rich in nutrient-dense vegetables, fruits, legumes, and whole grains. (3) Limit consumption of animal products and added sugar, which are known to affect your metabolism, increase your cholesterol, and clog your arteries. (4) Drink plenty of water. (5) Quit smoking, and avoid second-hand smoke as much as possible. (6) If you need to lose weight, lose weight, as directed by your doctor.

As logical as all this might sound, heart disease is still the number one killer of men and women in the United States, along with many other countries. Part of the problem is rooted in the food culture itself. For example, many Americans grew up on a "meat and potatoes" diet, one that also encouraged multiple glasses of milk and pancakes as part of a "hearty" breakfast. Even more than pancakes, bowls upon bowls of sweetened, processed, unhealthy cereal still remain the daily quick fix for breakfast and are even given to kids as a snack. Since this was considered the picture of a healthy diet at one time, it's hard for many to believe that these foods might be unhealthy.

A year ago, my husband was in Las Vegas when he sent me pictures of the Heart Attack Grill. That's the name of an American hamburger restaurant where diners don hospital gowns before indulging in "heart attack–inducing fare" such as the Bypass Burger. If that weren't enough, before one enters the restaurant there are scales on which incoming customers can weigh in. A flashing neon sign announces: "Over 350

Pounds Eats Free." Some of those people just short of the cutoff appeared disappointed to miss out on the deal!

If the threat of a heart attack is not enough to compel you to trade in your cheeseburgers and recliners for healthier food choices and a brisk walk, snowballing evidence that poor heart health is also bad for your brain might help do the trick. We need to keep our blood flowing to keep our bodies and brains full of life and longevity.

YOUR BRAIN IS A BUSY BEE

On top of eating healthily and keeping physically active, there is general consensus among scientists that exercising one's brain intellectually slows down aging and reduces the risk of cognitive impairment later in life.

New research boosts the "use it or lose it" theory about brainpower and staying mentally sharp by showing that people who retire at an early age have an increased risk of developing dementia. Of course there are retirement stories both ways. Some people have a great time after retiring. Others seem to go downhill physically or mentally shortly after their last day at work. Research shows that, on average, work seems to keep people active, socially connected, and mentally challenged, so much so that among nearly half a million people, those who delayed retirement by just a few years showed less risk of developing dementia in the years to come. For each additional year of work, risk of dementia went down by 3 percent.

This is not to say that you should work forever. Rather, the key is to keep yourself intellectually engaged throughout the course of your entire life. For example, a study of over four hundred community-residing seniors, most of whom were retirees, showed that those who regularly engaged in intellectual activity had a 54 percent reduced risk of cognitive decline as compared to those who did not.

So what qualifies as an "intellectual activity"? These activities can be anything from doing crossword puzzles and brainteasers to reading books and newspapers. Other options might be writing, playing music, joining a book club, or going to a show you enjoy. In fact, brain imaging studies show that lifelong participation in such activities slows down, and perhaps even prevents, any accumulation of Alzheimer's plaques, therefore protecting the brain against aging and dementia.

This brings us to a hot topic in the anti-aging field. In recent years, there has been an explosion of computer-based cognitive-training software, popularly known as "brain games." This online programming claims to make you smarter and improve your memory, while bumping up your IQ a few points at the same time. Claims like these can actually enrage quite a few scientists.

In 2014, the Stanford Center on Longevity and Berlin's Max Planck Institute for Human Development published a call to arms against the brain-training industry, signed by seventy-five of the best-known scientists in the field. In this consensus statement, the authors criticize companies for exaggerating claims and preying on the anxieties of elderly customers trying to stave off memory decline. Perhaps in response to increasing concerns such as these, the Federal Trade Commission (FTC) started paying more attention to online brain games companies. Just a few years later in 2016, the FTC took exemplary action against the company behind Lumosity, a well-known brain-training program. The company ended up paying a $2 million fine for engaging in "deceptive conduct," ergo false advertising, for claiming that their online games could delay cognitive impairment, memory loss, and Alzheimer's.

I'm often asked what I think about these brain-fitness products. To be honest, I have mixed feelings about them. On the one hand, some clinical trials show that cognitive training can improve performance in the elderly. For example, a study of almost three thousand elderly showed that participation in a brain-training program led to improved memory, reasoning, and processing speed after just a few weeks. These same participants continued to show above-average cognitive performance even five years after the intervention took place. This is one example of the kind of studies we often hear about in the news.

On the other hand, there are several trials with negative results or that report minimal improvements—and these are the studies that don't make the news. When we look at all the data as a whole, it turns out that this sort of cognitive training is only modestly effective at improving cognitive performance in older adults. In the end, just as with any drug or treatment with therapeutic claims, these products need rigorous testing in clinical trials and subsequent FDA approval before any conclusions about their efficacy is clear.

In the meantime, here's my advice. If an hour spent doing software drills, sitting alone in front of your computer or tablet, is an hour spent *instead of* walking, reading a book, or going to a show with your friends, then it's probably not worth it. If, however, you choose to play these brain games instead of sitting in bed or on the couch mindlessly watching TV, by all means, play brain games instead.

In this case, you might be surprised to learn that, among all the intellectual activities at our disposal, the human brain seems to actually have a favorite. It loves board games the most.

Several studies have identified playing board games as the intellectual activity most consistently linked with a reduced risk for dementia. In one example, a two-year-long study of four thousand people showed that those who regularly played board games had a 15 percent lower risk of dementia later in life as compared to nonplayers.

This makes sense, since playing board games is a highly stimulating activity. Far from merely being a source of entertainment, these games typically promote complex reasoning, planning, and attention, as well as memory skills. Plus, you are interacting with other people *and* are motivated to beat them. Some board games can be really challenging, such as chess or checkers. Card games are included in this group, proving to be as effective as board games when it comes to brain benefits. As anyone who's ever tried playing bridge would know, some card games can be real brain teasers.

As you might notice, all these games promote social interactions and often reinforce multigenerational bonding too. For many families, playing Scrabble on a rainy day makes for a special memory. In Italy it is quite common to find entire groups of retirees playing Briscola (a Mediterranean trick-taking card game) while sipping their *espresso*, grandchildren on their laps.

After all, we are social animals. A fairly big chunk of our brains—the limbic system—is all about loving and bonding, as much as playing. The feeling of being part of a group has always been a primary need for our race. Research shows that this need is in part motivated by the fact that people with a strong support system seem to live better and longer than others. As we have seen in chapter 9, having a sense of purpose and social connection can significantly increase longevity in the elderly and is an essential component of many cultures that show low dementia rates. A review of more than 300,000 participants shows that those elderly with stronger social networks have a 50 percent higher likelihood of living longer than those with fewer social ties or less satisfactory relationships.

Are introverts doomed? Not at all. As with so many things in life, it is the quality rather than the quantity of the relationships that really matters. A community-based study of over one thousand elderly showed that having a family you love is enough to stave off dementia, provided you connect with them happily and as often as possible. Married people, people living with others, or those who had children had a nearly 60 percent lower risk of dementia compared with people who lived alone or had no close social ties. In particular, parents with daily-to-weekly positive contact with their children had the lowest risk of all, while those who had relatives and friends but didn't see much of them, or felt these relationships were unfulfilling, showed the highest rates of cognitive decline.

Evidently, a loving brain lives a happier and longer life.

THE BRAIN'S BEAUTY SLEEP

Sleep, or a lack thereof, is the latest addition to the increasingly long list of lifestyle factors that can affect brain health. While a sound night's sleep has long been advised for a healthy body, it turns out the brain needs its sleep, too.

Experts agree that sleeping is crucial for memory consolidation and learning, and that poor sleep negatively affects these much-needed abilities. Without adequate sleep, your brain becomes foggy, your attention dwindles, and your memory stalls. This might not be news to anyone who has pulled an all-nighter cramming for a test only to find they couldn't recall most of the information the next day. Anyone who's experienced chronic sleep deprivation is very well aware of its effects. As a new mom, I have firsthand experience as to how serious an impact sleep deprivation can have on brain function.

Unfortunately, we are conditioned to think of sleep as a commodity, one that you must often forsake for other more pressing needs, say, a work deadline. Especially in the United States, needing sleep, sleeping a lot, or liking to sleep late are each associated with a lack of productivity, while people who are constantly on the go are applauded.

What many people don't realize is that a lack of sleep is a serious threat to the health of our brains, and might even deteriorate our cognitive function at large as well as increase our risk of Alzheimer's. In fact, one feature of sleep that most people don't recognize is its ability to *clean* the brain of harmful toxins, waste products, and damaging free radicals.

Only in recent years have scientists figured out how the brain's unique waste-removal technique actually functions. These studies revealed that whenever the brain needs to clean itself up, it employs the *glymphatic system*. With a series of pulses, this system literally bathes the brain's tissues with cerebrospinal fluid. The fluid in turn rushes in and travels throughout the brain and, acting a bit like the jets of a dishwasher, flushes away accumulated toxins and waste.

While many of us take our showers first thing in the morning, our highly unique brains prefer doing that at night. The glymphatic system is programmed to perk up and launch its activity just as we're about to sleep deeply. Research has found that in lab animals, brain clearing becomes ten times more active during sleep than wakefulness. It was during this time that harmful toxins like the amyloid proteins linked to Alzheimer's were flushed out of their brains. When the animals didn't get enough shut-eye, those same toxins built up night after night, damaging the brain as a result.

Brain imaging studies indicate that this might be the case in humans as well as animals. In some studies, older adults who slept less than five hours a night, or longer but fitfully, showed higher levels of Alzheimer's plaques in their brains than those who slept soundly for over seven hours. More work is needed to clarify whether poor sleep accelerates the buildup of plaques by hindering their removal, or whether plaque accumulation is a cause of poor sleep—or both. Either way, getting too little sleep or sleeping poorly is tied to an increased risk of mental deterioration.

So how long should we be sleeping for?

There is no magic number of hours of sleep that is right for each and every person. However, if research is any indication, we do need to give the brain adequate time to clean itself. Here's the catch. The brain's housekeeping activity takes place during a very specific sleep stage known as "deep sleep."

You might have noticed that your sleep is not uniform throughout the night. Each of us goes through sleep cycles that last about 90 to 110 minutes and include five separate stages of sleep. The first stage is actually falling asleep. The second stage is known as "light sleep," and is the brain's way of preparing to shut down. During the third and fourth stages, your brain is in a deep, or slow wave, sleep. It is during this stage that everything seems to fully come to a halt. Your muscles relax to the point of becoming inert. There are no eye movements. You are essentially cut off from the world. At this moment, you are sleeping a deep, dreamless sleep. This is the perfect chance for your brain to enjoy some much-deserved me time.

As your body reaches a deep stillness in this state, needing close to no supervision, your brain gets busy taking care of itself, washing away toxins and getting rid of all sorts of waste products. After a while, it is interrupted by stage five's rapid-eye movement (REM) sleep, during which we dream. But when REM sleep is over, this five-stage sleep cycle begins all over again, and soon enough, your brain will prepare itself for yet another shower.

If you sleep seven to eight hours per night, your brain will go through a number of these cycles. The first of these cycles will have the longest period of deep sleep and the shortest period of REM sleep. Later in the night, your REM periods will lengthen while your deep sleep stages decrease. If you want to make sure your brain has enough opportunities to clean itself, guard your sleep, especially during the first part of the night.

MOVE, LOVE, LAUGH, BE WELL

Even though diet is a powerful preventative against brain disease and cognitive impairment, diet alone is not enough. In fact, nothing alone is enough. As we've referenced earlier, synergy and a holistic perspective are the keys to continued health. It is time we learned to think about our bodies as a whole and about our lives as a combination of different sources of nourishment, where nourishment includes but is not limited to the foods we eat.

In addition to eating well, other forms of nourishment include how often we move or exercise our bodies, how connected we feel with our friends and family, how often we challenge ourselves intellectually, how content we are with our careers, and even how soundly we sleep. Each of these elements supports brain health, but when existing together with one another, they become greater than the sum of their parts. The extent to which we are able to incorporate all these healthy habits into our everyday lives determines just how healthy and durable our brains and bodies become.

And yet many people still find it hard to believe that leading a healthy lifestyle can boost our brainpower over the course of a lifetime and even ward off diseases like Alzheimer's. Has it actually been proven that such a lifestyle modifies Alzheimer's risk? Where are the clinical trials that demonstrate this causal relationship?

At long last, here they are.

A groundbreaking clinical trial published in 2015 showed that relatively easy lifestyle-based strategies to fight dementia, including diet, exercise, intellectual stimulation, and vascular risk management, were indeed successful in improving cognitive performance in older adults. Over as little as two years, the participants showed a 25 percent improvement in cognitive performance. The program was particularly effective at boosting people's ability to carry out complex tasks like remembering phone numbers and running errands efficiently, which improved by 83 percent. Even better, the speed at which they were able to conduct these various tasks improved by as much as 150 percent.

Except for sleep, this study succeeded in incorporating all known lifestyle ingredients for a healthy brain, providing important proof of a *causal* relationship between lifestyle and cognitive fitness. Research has finally begun to show that people who lead well-rounded, healthy lives with attention to these crucial, interactive elements are effectively improving the health of their brains and reducing their risk of dementia. With so many negative trials reported regarding the use of Alzheimer's drugs, these findings offer us the much-needed alternative we've been seeking. No longer out of reach, even the most skeptical among us is empowered with renewed hope as well as the motivation to do what's necessary to safeguard ourselves and thrive.

Now do you feel like dancing?

STEP 2

EATING FOR COGNITIVE POWER

11

A Holistic Approach to Brain Health

YOUR BRAIN'S FAVORITE MENU

We are now going to put the knowledge gained into practice and explore the *basic* guidelines for optimal brain nutrition. This chapter provides dietary and lifestyle recommendations for everyone who wishes to enhance their brainpower, improve their memory, and protect their cognitive skills. In addition, it assists those specifically interested in harnessing their diet to better optimize brain health, slow its aging, and minimize the risk of Alzheimer's. These recommendations are based on solid, scientific evidence regarding those nutrient combinations deemed essential and necessary for maintaining peak brain performance that we've seen in the previous chapters. In addition, they employ the latest key concepts from the fields of nutritional medicine, microbiome, and nutrigenomics research.

First and foremost, we must find ways to increase your intake of brain-essential nutrients, as these are critical to proper brain function. A rundown of the "superfoods" that best provide these brain-essential nutrients is included, along with several practical tips as to how best combine them to achieve continued brain health. As you embark on eating healthier to extend your life and enhance your mental well-being, remember to focus on the most important foods and add them to your daily meals however you can.

Also remember that it isn't only about increasing consumption of brain-healthy superfoods but also deliberately doing so while you *reduce* intake of foods that will instead harm your brain. Pay special attention to those also affecting heart health, such as processed, fried, and fatty foods. Be careful not to overeat red meat and dairy. If this sounds intimidating, don't worry. I'm here to make sure you won't have to deprive yourself to eat right for your brain. I have the "how to" when it comes to replacing brain-harmful foods with healthier and even more satisfying alternatives.

Moreover, the plan isn't just a matter of dieting so much as it is a matter of changing your lifestyle to ensure that the majority of the choices you make are supportive of your brain's health. As we talked about in earlier chapters, the best evidence to date shows that cognitive health in old age reflects the long-term effects of a healthy and engaging lifestyle. Physical activity, intellectual stimulation, social interactions, and good sleep are all part of the ensemble cast that works together to keep our brains active, vibrant, and alert over the course of a lifetime. To this end, I've included recommendations that address not only diet and nutrition but also those other areas of life known to directly contribute to the health and well-being of our brains.

THE MAIN COURSE: PLANT-BASED FOODS

As we saw in chapter 2, the brain's evolution was quite a lengthy affair that took place over the course of millions of years, as our ancestors emerged from the forests and gradually devised better and better strategies to feed themselves. What might have begun with daggers and bows eventually evolved into farming and agriculture. As a result, in the beginning, and for the longest extended span of time since our species began, our developing brains obtained their nourishment from a very specific and unpretentious diet. If we were to describe their culinary preferences, we could say that our early ancestors were raw vegans. Plant-based foods are what our brains first nourished themselves with and still need as their foundation of optimal health.

If you remember the typical centenarian diet, those who have mastered the secrets to a long, healthy, dementia-free life are actually 98 percent vegetarian. By and large, their diets rely heavily on fresh vegetables, fruits, grains, and legumes eaten on a daily basis. In addition, these are foods that are naturally low in calories while being packed with nutrients fitting our brains like a glove. From so many different perspectives, their very nature mirrors that of the human brain and does so in a way that is not equaled by any other food group. Brain-essential vitamins, minerals, good carbs, good fats, lean protein—plant food has it all. These foods are also the best source of antioxidants like vitamins C and E, beta-carotene, and selenium.

Leafy greens like spinach and Swiss chard, as well as fibrous vegetables such as broccoli, asparagus, and cabbage, are great sources of brain-essential nutrients. Citrus fruits, berries, and sweet potatoes are also chockful of goodness. Do you like avocados? They are a brain must. What about nuts and seeds? Almonds and Brazil nuts, flax, and chia are nothing but good for you. To give you a sense of their potency, just a handful of Brazil nuts contains 800 percent of the recommended daily dose of selenium, a major anti-aging mineral that is otherwise very difficult to obtain.

Additionally, there is an abundance of phytonutrients to consider. As the term itself implies, there is no food group that provides phytonutrients in amounts as plentiful as plants do. Many of these compounds are already powerful antioxidants, but in combination with the vitamins and minerals mentioned on the previous page, they act as an ideal brain-protective potion.

The digestive benefits that come from the high-fiber content in plant foods are another major plus. Dietary fiber is critical for the health of our guts and brains, not only on a daily basis, but even on a meal-by-meal basis. Vegetables are among the richest sources of fiber that exist, with grains, legumes, and berries coming in a close second. Last but not least for many of us, plant foods are rich in natural "no sugar added" glucose, and that can satisfy your brain's sweet tooth without harming your insulin levels at the same time.

Though people of all walks of life have different health and nutritional needs overall, one can't go wrong putting plant foods front and center. Am I asking you to become a vegetarian? No, but any way we can work more plant-based foods into our meals is a vote for future health. The goal is to eat vegetables with lunch *and* dinner, whole fruit at least once a day, and whole grains and legumes at least four times a week. As a rule of thumb, plant-based foods should take up the largest share of your plate.

Nature offers endless options to this end. Vegetables and fruit, legumes, whole grains, starches, not to mention nuts and seeds, are but a few of the options available. Within each of these choices, there's so much more than meets the eye. Vegetables alone include an astonishing number of varieties. Did you know there are as many as 150 kinds of cabbage? And hundreds of types of squash?

Yet the typical American diet remains fairly narrow when it comes to its daily vegetable consumption. According to the USDA, far and away the most popular vegetable in America is the white potato, followed by the tomato. This wouldn't be so bad if it weren't for the fact that these otherwise potentially healthy vegetables are usually consumed in the form of French fries and pizza. To top it off, we've chosen the most nutrient-empty vegetable of all, vapid iceberg lettuce, to crown our hamburgers.

Unfortunately, these foods won't help your brain at all.

In the next pages, we'll discuss how best to increase consumption of plant-based foods to specifically promote cognitive health and fitness. In addition, I'll share several of my brain-healthy secrets to eating right for your brain.

Here's the first: dandelion greens. I promised you'd hear more about these greens, didn't I? I came to love dandelion greens as a child. My *nonna* would often serve these greens as part of our Sunday meal. To this day, I distinctly remember how much lighter and clear-headed I would feel by the afternoon. *Nonna* used dandelion greens in a number of recipes, but in none more so than her staple spring side dish. She blanched the greens in boiling water, sometimes using the flowers, too, and then served them with freshly squeezed lemon juice and our favorite extra-virgin olive oil, bought directly from a neighboring Tuscan farmer. The realization that this dish brims with brain-essential nutrients was the start of my lifelong commitment to neuro-nutrition (as well as my renewed respect for the lowly dandelion).

If dandelion greens aren't on your radar, they definitely should be. These leafy greens are a main staple of Mediterranean cuisine, and in addition to being a delicious food and having medicinal properties, they can even be grown at home and put to use. Believe it or not, they are packed with just about every nutrient your brain craves. Although they are not the telltale orange color, these greens are a very rich source of vitamin C and beta-carotene. They are also abundant in vitamins E, K, choline, folate, and B6, with a rich mineral and fiber content. What surprises a lot of people is that they have quite a bit of protein for a plant. One chopped cup boasts 1.5 grams of lean protein containing all the essential amino acids. If that weren't enough, their distinctive, slightly bitter flavor is a sign of these greens' ability to nourish your friendly gut bacteria. And as you will recall from the previous chapter, dandelion greens contain molecules that boost our cardiovascular system's rejuvenating properties. Where can you find these exceptional greens? On FreshDirect or AmazonFresh, for starters. Probably in your garden, too.

Here's another brain-healthy secret to boost your intake of healthful veggies, grains, and legumes: Buddha bowls. There's no better way to eat dandelion greens, or any vegetable for that matter, than as part of a Buddha bowl. Sometimes referred to as glory or hippie bowls, Buddha bowls are hearty, filling dishes made of raw or roasted vegetables, legumes such as beans and lentils, and whole grains like spelt or brown rice. A Buddha bowl is so full of all things good that when served, it has the appearance of a rounded belly (much like the belly of a Buddha). Depending upon the recipe you choose to follow, the dish can contain a rainbow of ingredients. They can also include toppings like nuts and seeds—and come with incredibly flavorful dressings, like my Maple Tahini Dressing (chapter 16). And the best part is, Buddha bowls are simple to make and jam-packed with filling nutrients and vitamins that nourish and protect your brain. Since it can take a little while to prepare all the ingredients, I typically make a large batch of everything (rice, spelt, buckwheat, steamed dandelion greens, roasted veggies—even the dressing) and keep them refrigerated in airtight glass containers. Some of my favorite recipes are included in chapter 16, and many more are available at www.lisamosconi.com.

THE RIGHT FATS, IN THE RIGHT AMOUNTS

Increase your intake of brain-healthy fats while limiting your intake of the artery-clogging ones and we're off to a good start. All fat, however, healthy or not, is rich for the body and should be consumed in moderation. A key strategy here is to limit the total intake of fats to those that are actually beneficial to the brain and forgo those that are not. Achieve this and it will help your entire body in the process.

These healthy fats include omega-3 PUFAs and especially DHA—that rare fat that abounds in marine and fish oils. As you will remember, there is a consensus that eating high-quality fish and shellfish is not only good for your brain, but also reduces the risk of memory loss and dementia. Besides being loaded with omega-3s, fish is an excellent source of complete protein and vitamin B12, critical players in the overall health of your nervous system. Wild-caught fish is the best source of DHA. Some of my favorites are Alaskan salmon, mackerel, blue fish, sardines, and anchovies.

Most longevity diets recommend eating fish at least once a week. We will follow suit and raise the bar to two or three times a week.* The trick here is to focus on high-quality fish, as well as pairing it with those foods that enhance its qualities, such as select herbs and, lest we forget, even a glass of wine. One example could be your choice of fish, roasted with lemon, herbs, and sea salt, or covered in crushed pistachios as a special treat. And then there's my secret weapon—caviar.

Considered by many to be a luxury gourmet food, black caviar consists of salt-cured fish eggs from sturgeon. It takes no more than 2 to 3 teaspoons of caviar to reach your brain-healthy DHA *and* choline dose for the day. Of course, the downside is that caviar can be quite expensive. My favorite alternative is salmon roe, which contains almost the same amount of DHA while costing a third as much as black caviar. In addition to being incredibly rich in brain-healthy fats, salmon roe contains high levels of antioxidants, like vitamins C, E, and selenium, combined with a nice dose of B-complex vitamins. It is also high in protein. Just one ounce of salmon roe contains 6 grams of protein chockful of essential amino acids. Salmon roe is very versatile and can be "dressed up or down," depending on the occasion. Add a couple of teaspoons on your favorite sushi roll, or sprinkle some on a rye bread *tartine* to make a snack, or on whole-grain toast over Greek yogurt to serve as a fancy hors d'oeuvre.

When you're not eating fish, you still need to make sure you're hit-

* Make sure that you don't overdo it with large fish such as tuna and shark, which tend to be high in mercury. Pregnant women should always avoid fish with high mercury levels.

ting your omega-3s. Several nuts and seeds can help. My favorites are almonds, walnuts, flaxseeds, chia, and hemp, which I routinely add to smoothies, soups, and salads.

It is also useful to swap omega-6-rich oils for those that contain omega-3s instead. Flaxseed oil tops the list of omega-3-rich oils, with over 7 grams of omega-3 per tablespoon. Oils and products that are rich in omega-6 need to be consumed *in moderation only* and these include grapeseed, sunflower, corn, soybean, sesame, and peanut oils.

Additionally, heart-healthy monounsaturated fats, such as those found in macadamia nuts and high-fat fruit like olives or avocado, should also replace less healthy fats. Olive oil is perhaps the most publicized and widely used source of these good fats, especially the cold-pressed, extra-virgin variety made from the first pressing of the olives. It is now believed that regular consumption of extra-virgin olive oil is a prime reason for the positive aspects of the Mediterranean diet, thanks to its unique high-antioxidant blend.

There is a type of fat that is unequivocally harmful for you and that needs to be completely eliminated from your diet first and foremost. I'm talking about trans fats. Remember how these fats raise your cholesterol levels, producing painful inflammation throughout your entire body? On top of that, food that contains trans fats typically also contains poisonous metals, emulsifiers, chemical sweeteners, and artificial coloring, which all work to shut down your brain, heart, and microbiome.

Trans fats are typically lurking in processed foods. Commercial doughnuts, cookies, crackers, muffins, pies, cakes, Cool Whip–like creams, processed cheeses, and candy—these are just some of the foods that contain trans fats. But processed meats like bologna, salami, corned beef, and pastrami are also culprits. Vacuum-packed mozzarella? Canned cheese that squirts? *Arrivederci!* These, too, are shot through with trans fats. It is crucial to limit how frequently you eat these foods until you're down to never. Go for organic sources instead. They aren't much more expensive, and they have the added bonus of containing healthier fat and far less sugar to boot. A fresh, homemade apple pie is clearly more delicious than an industrialized, preserved one that can live endlessly on a supermarket shelf. Don't bake? Find someone who does. You and your health, as well as the health of your family, are worth it.

It is also important to limit fried foods and baked goods made with shortening or partially hydrogenated vegetable oils—which is the absolute norm at every single fast-food chain in spite of their sometimes marketing such products as "natural" or "healthy choices." These include anything from French fries to fried chicken, but also items such as fried mozzarella and vegetable sticks, anything battered, and nearly all candies and cookies. Do you love chips? Make your own. Sweet potatoes cooked to a crisp in coconut oil are a real treat, and far more nutritious than any bleached, deep-fried fakery you can find at the local fast-food joint or convenience store.

As you reduce your consumption of processed foods, you'll be reducing saturated fat in your diet in the process. Saturated fats, especially those from natural sources, don't need to be entirely eliminated but should be significantly reduced. Since the body burns saturated fat for energy, just enough of this fat needs to be provided for all bodily functions to occur smoothly. At the same time, we need to stop overconsuming these fats for three very good reasons: to protect our brains against aging, to avoid weight gain, and to lower the risk of heart disease. The truth is, while the general public might be conflicted over which fat is good and bad for you, scientists go to the trouble of doing the math. In the long run, excessive saturated fats are bad for you.

As previously mentioned, there are different types of saturated fats and some are better for you than others. For instance, some vegetable oils, like coconut oil, are excellent sources of a particular type of saturated fat, the so-called medium-chain triglycerides. There is increasing evidence that these oils don't have a negative effect on cholesterol, and can actually help to lower the risk of atherosclerosis and heart disease, therefore reducing the risk of dementia, too. They are also naturally cholesterol-free, which is a plus if cholesterol is an issue for you. Additionally, medium-chain triglycerides are a good source of ketone bodies, the brain's backup energy source in times of scarcity and fasting. However, since we will be providing plenty of healthy glucose to our hungry brains, these high-fat foods are not really necessary for energetic consumption. Besides, they shouldn't be consumed instead of omega-3-rich oils, which are far more crucial for a healthy brain. So use them judiciously, as described in the next chapter.

Saturated fats from animal products are a different story. Remember how centenarians all around the world tend to eat less meat and dairy—and often only at community celebrations? When eaten in excess, these foods are potentially dangerous, as they are rich in triglycerides and cholesterol, as well as the omega-6s that actually compete with the omega-3s for entry into the brain. As a result, they can boost inflammation and raise cholesterol levels, in turn increasing the risk of vascular damage.

I am not suggesting you give up your meat and cheese *entirely*. What I am talking about is portion size. Many people chow down on double and even triple portions of steaks and burgers—often loaded with processed cheese on top. If you need to eat a pound of meat to feel satisfied, this comes under the category of overindulging and it's a problem. Use your hands to guide you: one serving of meat of any kind should mirror the size of the palm of your hand or a deck of cards, equaling approximately 2 to 3 ounces. One serving of cheese is about the length (width and depth) of your index finger, equaling about 1 ounce.

Frequency is also important. Red meat and pork should not be consumed more than once a week. Focus on lean, rather than fatty, cuts, and if you're eating chicken, get rid of the skin. Cheese should also be restricted to no more than once or twice a week. Milk, on the other hand, can be a good source of many essential nutrients. When you drink milk, or use it in your cooking, focus on organic grass-fed milk. Also, there's no need to drink a quart of milk. Think of milk as liquid food. A small cup is plenty, especially if it's in combination with other foods.

Yogurt is the exception to our careful dairy rules. Yogurt is an excellent source of brain-essential nutrients and probiotics; consequently, you can have a cup every day. Eating yogurt on a regular basis is crucial to maintaining optimal GI functions, which in turn support brain health.

Finally, a word about eggs. Eggs are a favorite breakfast choice in America and many people eat them daily. Research has shown that even though eggs are not as bad for you as previously thought, we shouldn't be overdoing it. I typically recommend eating two to three eggs throughout the week, either scrambled or poached, or as part of omelettes and home-baked goods like my Blueberry Banana Muffins and Banana Almond Pancakes (www.lisamosconi.com).

Hold on until chapter 12 for a discussion of the different types of egg, meat, and dairy products.

FOR DESSERT, THINK GLUCOSE

We need to increase our consumption of "good carbs" while decreasing the bad ones. When we look around at the typical Western diet, in addition to the consistent consumption of poor-quality meat, the one thing that seems more evident is the rampant consumption of refined white sugar via fast-food meals, processed products at our tables, and unhealthy snacks available around every corner. Not to mention the rows of artificial food that passes for breakfast cereal in our grocery store aisles, alongside cheap pastries, cookies, and energy bars—all sugar- and chemical-laden. Not even vegans are safe. Although "vegan foods" should be the epitome of healthy, more often than not they are packed with an ungodly amount of hidden sugar, making the consumers not much better off than their meat-eating counterparts.

Nowadays, more and more doctors attend to sugar consumption as much as they do to unhealthy fats with regard to heart disease in particular. However, these sugary foods will do more harm than good to your brain as well. These are the "bad carbs."

As the public becomes increasingly aware of just how bad refined sugar is for the body, the tendency has been to replace white sugar with artificial sweeteners such as aspartame (NutraSweet, Equal), sucralose (Splenda), acesulfame potassium, or Ace K (Sunett and Sweet One), and saccharin (Sweet'N Low). These sweeteners have come under scrutiny due to damaging side effects that range from headaches and migraines to impaired liver and kidney function, not to mention mood disorders.

Fortunately, there's no need to ingest toxins such as refined sugar, high fructose corn syrup, and artificial sweeteners. There are plenty of healthier and more natural sweeteners at our disposal.

Remember, your brain runs on glucose. Therefore, as far as your brain is concerned, "good carbs" equal the *glucose-rich foods* described in chapter 6. These include not only raw honey and maple syrup but also coconut sugar, brown rice syrup, yacón syrup, blackstrap molasses, stevia, fruit purees (aka fruit butters), and even fruit like grapes and vegetables like beetroot. You might be surprised to discover that these natural sweeteners come with an added bonus—they increase your intake of anti-aging antioxidants in amounts similar to that of consuming berries and nuts. Give these a try to discover which natural sweetener you like best. More specific recommendations tailored to your specific diet are included in the next chapters.

Some complex carbs, particularly whole wheat, sorghum pasta, wild rice, and sweet potatoes, are also rich in glucose as well as fiber, vitamins, and minerals. This particular nutrient combination provides sustained brain energy for longer periods of time, making these foods ideal lunch options. Many people enjoy their cereal for breakfast. If this is you, I recommend wholesome, unsweetened, minimally processed, 100 percent whole-grain cereals, free of artificial coloring and synthetic vitamins or minerals. These cereals won't come in brightly colored boxes with catchy brand names. Rather, you'll find them in nondescript transparent bags with no more than a small label. Great examples are steel-cut oats, puffed brown rice, and buckwheat porridge. Just sweeten them yourself by adding a little honey or maple syrup, or topping your bowl with fresh fruit.

That being said, I confess to having a sweet tooth and appreciate how staying clear of dessert can be a real challenge. For me, that wasn't much of a problem when I lived in Italy, but after I moved to New York, I started having sugar cravings. As soon as I ended my lunch, I would find myself reaching for a cookie or a piece of chocolate, mostly in response to a sudden drop in energy. Those rare days when I couldn't lay my hands on either, I would find myself in the worst mood imaginable—and my performance suffered as well. I know that I'm not alone and that many people can relate to craving something sweet after a meal. Making things worse, we beat ourselves up after indulging.

The thing is, sugar cravings often result from a poor diet. Unfortunately, most American meals are full of refined sugars that upset your blood sugar levels, causing you to only want more. On close examination, the treats I used to have in Italy were most often homemade with fresh, organic ingredients *and* low in refined sugar. They had a very different effect on my body than the supermarket chocolate chip cookies I got used to in my new American home. It was quite a process kicking the habit. I had to learn to read labels, figure out ingredients, and focus on natural sources of sugar rather than commercial ones. But it *did* pay off. As I got smarter about my diet, I stopped having sugar cravings (or the sugar blues) and I lost whatever weight I had gained without trying so hard. If you too need to "de-sugar," the diet plan outlined in Step 2: Eating for Cognitive Power will help you achieve this goal while fueling your brain at the same time.

Making sure that your plate is filled with foods that are healthy and satisfying will in and of itself reduce the need for sugary desserts, soda, and extra coffee. And when you do need a little pick-me-up or are in the mood for dessert, there is no need to deprive yourself. Just be mindful of what you put in your body and how often. Many of my favorite brain-healthy desserts, snacks, and occasional treats, like Chocolate Almond Power Bites, Chocolate Blueberry Ice Cream, and Raffaello Coconut Butter Balls, to mention a few, are available on my blog (www.lisa mosconi.com). These recipes are all full of glucose and low in calories and glycemic load. Hope you'll like them as much as I do!

Finally, when in doubt, eat chocolate.

BYE-BYE, HERSHEY'S

Of all the treats available to mankind, chocolate has been one of the most craved foods in the world since ancient times. Already the Aztecs and Mayans regarded chocolate as "the food of the gods," consuming it with great reverence. Back then, "chocolate" consisted of an exotic, bitter beverage made of fermented, roasted cacao beans that were ground into a paste, mixed with water and exotic spices, and sweetened with honey. When consumed in its purest form, chocolate remains a powerful superfood with impressive health benefits. For this to be the case, however, you have to be willing to give up the milk and sugar that makes up most commercial chocolates. Real cacao is bitter, thanks to the action of hundreds of polyphenols, busy delivering health benefits. Unfortunately, the vast majority of the chocolate many of us have eaten throughout our lives is in the form of milk chocolate, white chocolate, and confectionary chocolate or candy, all containing only trace amounts of healthful cacao. What they do contain are extraordinary amounts of sugar, fat, and additives. For example, the typical Hershey's bar (milk chocolate) contains as much as 16 grams of sugar per 1 ounce of chocolate.

Dark chocolate, on the other hand, is low in sugar and rich in antioxidant flavonoids and minerals like magnesium and potassium. It also brims with theobromine, with positive effects on blood circulation and possibly on LDL ("bad") cholesterol. The key is to focus on high-quality dark chocolate with at least 65 percent cocoa and little or no added sugar. Keep in mind that different brands use different proportions of cacao, sugar, and cacao butter, so they are not all equal sugar-wise. For example, 1 ounce of 70 percent Lindt Excellence chocolate contains almost 10 grams of sugar, whereas 1 ounce of 74 percent Dagoba Organic chocolate contains only 7.5 grams. My all-time favorite, Lindt Lindor dark chocolate truffles, are each half an ounce of mouthwatering lusciousness and but 5 grams of sugar. On those days when I really just want to indulge, I'll go for the Lindt truffle. See how many options there are?

THIRSTY?

Drink water. This is my personal brain-healthy mantra.

Drinking *clean* water is essential in promoting hydration, restoring balance, and powering all sorts of cellular activity throughout your entire body. And yet most people don't drink nearly enough. A very common reason is that many Americans report not liking the taste of water, or perhaps it's because it barely has a taste at all. When I first heard this, I had trouble understanding what they meant. Since water has no particular flavor, how can it offend you? Eventually I came to understand that for people who've grown up drinking soda, milk, and fruit juice as their primary sources of fluids, drinking plain water just isn't much of a thrill.

Of course, there are alternatives. Herbal tea is a great start. Drinking herbal tea is a good way to stay hydrated and is a wonderful additional source of vitamins and minerals. Plus, there's the social aspect of sharing tea with friends that can make teatime count as an enjoyable, brain-loving activity. There are endless herbs and blends to choose from. Some of my favorites are soothing peppermint, calming rose and chamomile, cleansing rose hip, and of course anti-aging ginseng, ginger, and lemongrass. In the summer, these same hot teas are delicious served iced.

Fruit-infused water is another smart way to add flavor and nutrients as you hydrate (while helping you ditch the soda can on the way). Also known as detox or spa water, these beverages can be any combination of fruit, vegetables, herbs, and spices immersed in water to flavor it. Infused water is delicious without added sugar and has no calories, making it a powerful tool in your efforts to achieve better health. Several other options are available at www.lisamosconi.com.

To give you a sense of what these waters are all about, let me tell you how I make my Spicy Raspberry and Orange Water. You'll need a cup of raspberries with a thinly sliced orange, two cucumbers (also thinly sliced, skin still on), a handful of fresh mint leaves, and two cinnamon sticks. All you have to do is mix the ingredients in a large pitcher, add 1 gallon of spring water, and allow it to steep in the fridge overnight. You can add additional fruits and herbs if you prefer a stronger taste. When you're ready for your drink, add a cup of ice to the pitcher (I also add half a glass of aloe juice) and serve immediately.

Green juices and smoothies are another excellent strategy to increase your fluid intake throughout the day. Many people swear by cold-pressed green juice first thing in the morning, while others look on with suspicion at their bright green hue. Personally, I have learned to appreciate the therapeutic properties of these beverages, especially some smoothies. By smoothies, I am not referring to those milkshake-like beverages loaded with sweetened fruit juice, sugar, processed milk, and even ice cream. I'm talking about whole-food smoothies that are made of organic fresh fruit, vegetables, nuts and seeds, and plenty of water. While these drinks alone won't fix a poor diet, and are not meant to replace water, they are an easy, fast way to incorporate more fruit and vegetables as well as more brain-essential nutrients into your daily drill. Several recipes are included in the next chapters.

Another trick to staying hydrated is to reduce the amount of coffee you drink. Do you ever get thirsty after you drink coffee? Do you drink several cups a day? Drinking too much coffee might lead to unwanted side effects such as dehydration, palpitations, and disturbed sleep. Nonetheless, moderate coffee drinking in midlife might be protective against dementia later in life. What to do? One of the keys to smart coffee consumption is *how* you make your coffee. For example, *espresso* typically has over five times the amount of antioxidants, along with more caffeine than boiled or filtered coffee. In addition, different varieties of the coffee bean itself, along with its processing, produce different caffeine contents. So depending upon which preparation method and coffee you use, stick to either one demitasse cup of espresso a day or two cups of freshly brewed organic *Americano* (the classic, taller cup of American coffee).

While I most definitely love my espresso, it's great to have options. Cacao tea is by far my favorite coffee substitute. Cacao is an effective mood enhancer, an excellent antioxidant source, an energy booster, and a comfort food all wrapped up in one. Cacao tea will give you a nice hit of energy without the jitters and the crash that can sometimes affect coffee lovers—and it tastes delicious. I take my cacao tea black and unsweetened, but you can add a little bit of stevia or raw honey if you prefer. You can also make it creamy by replacing the water with almond or hazelnut milk.

Yerba maté tea, gunpowder green tea, and matcha tea are other coffee-kicking alternatives that are rich in antioxidants. Matcha is a fine-ground green tea powder that dissolves in water, making it a great iced tea, too. But if it's the actual taste of coffee you crave rather than the energy itself, try dandelion tea. Believe it or not, for centuries my beloved dandelion greens have been dried and roasted to make tea all across Europe and Asia. It's easy: let the dried greens steep for ten minutes in boiling water, then add a little milk and honey to taste.

A GLASS OF RED, PLEASE

It's pretty straightforward: if you drink alcohol, drink it in moderation, always on a full stomach, and choose red wine over other varieties as much as possible. Red wine has long been known to protect our brains from harm while uplifting the heart. As Italy's dear Galileo Galilei said, "Wine is sunlight, held together by water." While both white and red grapes contain resveratrol, the antioxidant compound that gives wine its good reputation, red grapes have a much higher content of it.

Alcohol is a very personal choice. If you are male, shoot for up to two 5-ounce glasses a day. If you are female, one 5-ounce glass of wine a day is enough. Here again, go for quality over quantity (organic wines are best), and find what works for you.

If you don't like wine, there are other choices. For example, organic pomegranate juice is almost as rich in antioxidants as your average red wine. Grape juice and prune juice are also valid alternatives.

1+1=3

A major problem with nutrition studies is that, typically, *one* nutrient is studied at a time. Of course, this is easier to handle in research settings, since it reduces the number of variables that must be accounted for. However, the risk is falling into the trap of oversimplifying these nutrients into one of two categories: good or bad.

Originally, the one-diet-fits-all approach seemed to work when the goal was all about remedying a specific vitamin deficiency. However, this only encouraged scientists to feel satisfied with how single nutrients (rather than nutrients working in tandem) affected our health. This somewhat outdated attitude still permeates both medical and nutritional practices, leading many health professionals to lead witch hunts against the latest "bad guy" nutrient that, once unmasked, must be fully eliminated from our diets. As all the confusion around cholesterol has taught us, this practice can make for a lot of tail chasing.

On top of that, this tendency completely disregards the fact that nutrients work in concert, not in isolation, and it's this teamwork that works to ensure optimal health. For example, let's look at lemons. In your mind, picture biting into a tart, tangy lemon. Is your mouth watering, mentally as well as physically? Just *thinking* about biting into a lemon will kick your senses up a notch.

Now imagine popping a vitamin C capsule into your mouth. Most likely, you'll experience nothing. We've extracted the vitamins out of the lemons, but the lemon has a lot more to it. Citric acid, minerals like iron and potassium, B-complex vitamins, and an array of phytonutrients such as *hesperetin*, *naringin*, and *naringenin* all interact with one another to ensure each lemon is as nutritionally effective as it can be. A pill can provide as much vitamin C as the lemon, but it's incapable of providing the full experience that activates your mind and body: literally, 1+1=3—a concept traditionally overlooked by Western medicine.

Based on this realization, a wave of new studies has sprung up to take a smarter look into what we call *nutrient synergy*. My own work shows that the combination of several nutrients, such as omega-3 PUFAs, B vitamins, and antioxidants like vitamins C and E, is particularly effective at protecting memory and mental sharpness. These studies also showed that just like certain combinations of foods and their nutrients favor our health, other combinations do the opposite. This is true of the unhealthy alliance between trans fat, saturated fat, cholesterol, refined white sugar, and sodium. This combo is especially bad news for your brain. Research shows that people who routinely consume these nutrients together, by eating a fairly regular diet of sweets, fried foods, processed foods, and high-fat meat and dairy, exhibit more pronounced brain shrinkage, poorer cognitive performance, and an increased risk of Alzheimer's as compared with those who don't. Similar results were already presenting themselves in participants as young as twenty-five years of age, only underlining that these foods have dangerous effects on your brain's ability to do its job, regardless of your age.

Ultimately, there is no single food that can provide all the nutrients necessary to fully support your brain. A variety of different foods is much better suited to provide you with all the necessary ingredients your brain needs on a daily basis. It's as simple as that.

There are many examples of how research findings on nutrient synergies lend themselves to flavorful, delicious dishes while at the same time boosting the nutritional value of your meals. For starters, my Grilled Salmon in Ginger Garlic Marinade (chapter 16) is an excellent source of brain-essential nutrients known to reduce brain shrinkage and improve metabolic activity in people of all ages. Additionally, it turns out that absorption of antioxidant vitamins is enhanced by high-fat vegetable products like extra-virgin olive oil, making my Nonna's Dandelion Greens with Lemon Juice and Extra-Virgin Olive Oil an ideal accompaniment to the salmon (also in chapter 16). Vitamin C enhances the body's absorption of iron when these foods are eaten at the same time, which is why I make a point of drizzling the greens with freshly squeezed lemon juice.

All the recipes and meal plans included in this book are based on maximizing nutrient synergies while providing you with all the necessary brain-healthy nutrients you need on a daily basis. Plus, they taste great, too.

12

Be Mindful of Quality Over Quantity

YOU'RE WORTH IT

The focus of our brain-healthy diet should be on whole, unadulterated foods. Ideally, we'd shop the farmers' market every day for fresh, local foods and make all our meals from scratch. In reality, we often live far from our food sources, have to buy foods we can store on shelves, and barely have time to eat dinner, much less prepare it. On top of that, if we don't find some ways around it, eating healthy can be expensive.

If you are concerned about spending too much money eating organic, you are not alone. Thankfully, there are several things you can do to increase your intake of wholesome, high-quality foods without breaking the bank. A good amount of information is available online, but here are some tips that I've been using consistently for years that have saved me a lot of money.

First, online organic food retailers are a tremendous help. There are several websites like Thrive Market, Vitacost, and even Amazon that offer Whole Foods–quality products at Costco prices. As long as you're buying nonperishable products, you can find a bit of everything, from canned chickpeas to wild rice, gluten-free rolled oats and flours, healthy cooking oils, and all sorts of non-GMO nuts and seeds. It's a great way to source tuna fish, anchovies, and sardines, and even baby food. In addition, check the websites of your favorite companies for coupons and special promotions. Each of these online stores often offers easy-to-access discounts to further sweeten the deal. The food sections at store chains such as Marshalls, Home Goods, and T.J.Maxx are also excellent sources for scoring high-quality organic products at discounted prices. Stock up on these new main staples to replace the old low-quality items in your pantry with all sorts of long-lasting but safe goodies, and save money at the same time. Surrounding yourself with these options will prevent you from habitually turning to outdated food habits.

With respect to fresh produce, my number one recommendation for quality is to stay clear of conventionally grown, genetically modified (GMO or GE) products. How do you distinguish a waxed piece of GMO fruit versus the real thing? A simple way of doing this is to check out the stickers attached to fruit and many vegetables. By reading the code on the sticker, you'll be able to tell which of these three categories the fruit belongs to: GMO; conventionally produced with chemical fertilizers, fungicides, or herbicides; or organically grown. Here are the basics of what you should look for. If there are only four numbers on your apple (e.g., 4131), this means that the produce was grown with the use of the above-mentioned chemicals. If there are five numbers and the first is an 8, that's the code for GMO produce (e.g., 84131). If there are five numbers and the first is a 9, (e.g., 94131) the produce was grown organically and is safe to eat. By the way, organic fruit is not always picture-perfect, but it will be full of flavor and deliver on nutrient value.

The Dirty Dozen and Clean 15

Unfortunately, depending on where it's coming from, organic fresh produce can be expensive. A good money-saving strategy is to take a look at the Dirty Dozen and the Clean 15 lists (Box 1) below. The Dirty Dozen represents the twelve most contaminated and pesticides-laden fruits and vegetables on the market. It's important to invest in the organic versions of these foods to avoid ingesting toxic loads of chemicals along with your meals. On the other hand, the Clean 15 is the safest produce that one can eat in their non-organic forms, since they are not sprayed as heavily with pesticides. Other foods fall somewhere in between, so choose organic whenever you can.

DIRTY DOZEN	CLEAN 15
Apples, celery, tomatoes, cucumbers, grapes, nectarines, peaches, potatoes, spinach, strawberries, blueberries, bell peppers	Onions, avocado, sweet corn, pineapple, mango, sweet peas, eggplant, cauliflower, asparagus, kiwi, cabbage, watermelon, grapefruit, sweet potatoes, cantaloupe

Box 1. Dirty Dozen vs. Clean 15

How about a few more tricks? Try not to buy prewashed and ready-to-eat fruits and vegetables, as they will cost you twice as much. Buy local produce when in season and freeze it to have at your disposal when it's out of season. You can also track down a farmers' market near you through LocalHarvest.org or the USDA. Get to know your local farmers, create personal relationships, and don't be shy about negotiating prices. Another good move is to be the last person to leave the farmers' market. Farmers are likely to cut their prices at the end of the day so they don't have to lug their produce back to the farm. You can also buy a share in a community-supported agriculture (CSA) program. In exchange for contributing to a local farm's operating expenses, you'll get a weekly box of fresh fruits and vegetables.

The Truth About Soy

Soy is a plant-based food that deserves special mention because it has become one of the most controversial foods on the planet. You might find it promoted as a protein-dense superfood one minute and blacklisted as a cancer-inducing poison the next. As we strive to reduce our overconsumption of animal products, knowing whether or not to consume soy becomes particularly relevant, given its potential as an alternate source of lean protein. Many have used soy as a meat alternative for decades. In fact, soy first became popular due to the Okinawan centenarians largely preferring it over meat in their diet. Since the Okinawans are one of our best examples of healthy centenarians, we need to take a closer look.

Are soy products such as tofu, tempeh, and soy milk acceptable . . . or not?

As we've discovered with regard to so many foods, soy's verdict depends upon the *type* of soy you're eating and how much of it. There are basically two kinds of soy—the kind eaten in Japan and the kind eaten throughout much of the Western world. Soy products in the United States today are made up of 90 percent genetically modified soybeans and are rife with pesticides and preservatives. As such, they can cause allergies, intolerances, and even systemic inflammation. Also, soy contains molecules called *isoflavones* that are known to influence our estrogen levels. Eating too much soy can indeed impact estrogen levels, possibly creating hormone imbalances in your body. Some people are more sensitive to this than others and have to avoid soy products entirely.

As a result, while the soy industry would have you believe that eating soy products is a smart move for your health, the GMO-laden, highly processed soy found in commercial tofu, soy milk products, and tempeh is far from a health food and closer to a health hazard. However, it isn't an occasional miso soup that will make you sick so much as the additional soy that you unknowingly eat from a multitude of unexpected sources. Soy is added to as many as twelve thousand food products in the United States, from common breakfast cereals and energy bars to snack foods and pastas. If you're not paying attention to labels, you might be surprised to find that soy is found in the *majority* of products on your supermarket shelves, mainly in the form of soybean oil. In addition, isolated soy proteins are widely used as emulsifiers to lend moisture to a product's texture, and are also often added to drinks such as lattes and frappés to emulate a creamy consistency.

So what type of soy is good for you? The soy used in Japan, which is almost always consumed in its organic form. Additionally, soy in Japan is often *fermented*, as in the case of miso (as in your miso soup), tempeh, and natto (a traditional Japanese food made of fermented soy beans). This is the soy that merits consideration as a health food and is the *only* soy actually worth eating. Organic, fresh soybeans and edamame contain all essential amino acids along with healthy amounts of the PUFAs your brain is so fond of. They are also packed with iron, fiber, and minerals such as magnesium, potassium, copper, and manganese. Freshly made tofu is an excellent source of brain-essential nutrients, too. During a visit to Kyoto, I was lucky enough to be presented with a subtly earthy, silky custard in a traditional restaurant—and was totally won over by genuine tofu made of traditional soy. The milk is equally healthy and delicious. Though fairly expensive, you can find both at health food stores like Whole Foods. Keep in mind that the Japanese don't overeat their tofu. A typical serving of tofu is the size of a golf ball. Unfortunately, in the United States, most soy products like tempeh contain adulterated soy and are highly processed and should definitely be avoided.

Fish, Meat, Eggs, and Dairy

Animal products are more of a challenge, both in terms of quality and convenience. Fish in particular, a brain must, can be tricky because you want to eat *good* fish without having to take out a loan. With respect to quality, my personal suggestion is to choose wild-caught over farm-raised fish whenever possible. You want your fish to be wild or fresh caught so as to avoid any pollutants, pesticides, and antibiotics your dinner might have already ingested. When first hearing this, many shake their heads and think, "I can't afford that."

I faced this same challenge when I moved to New York on a meager student salary and couldn't find a decent piece of fish in my local supermarkets. This eventually led to my devising a few effective strategies. To begin with, although it is true that high-quality salmon is more expensive than the conventionally raised unhealthy ones, it takes as little as 2 to 3 ounces to meet your brain's needs for the day. That's about the size of two pieces of sashimi. If you feel like you need a bigger portion, fresh mackerel, cod, and halibut are inexpensive choices that top the list of fish rich in brain-building DHA. Buying frozen fish is another option. Frozen wild-caught fish still beats fresh farm-raised fish any day, both in terms of nutritional quality and safety. Believe it or not, you can buy a pound of wild Alaskan salmon online for as little as $12 (clean and cut) and then thaw it as needed. Wild anchovies and sardines can easily be bought online for less than $2 a can. My beloved salmon roe isn't cheap, I'll give you that, but you can buy it by the pound and for a lot less money at any Russian deli or online. To be clear, a quarter pound will last you a week or longer. Remember that, as far as your brain is concerned, 2 tablespoons of salmon roe are worth 30 pieces of chicken!

Now let's tackle the meat question. Whatever you do, do not skimp here. Some consumers know to look for specific qualifications when it comes to shopping for meat, such as organic, free-range, grass-fed, pastured, hormone-free, and cage-free. While you might think these labels are interchangeable or unimportant, they're actually not. For example, did you know that free-range chickens don't necessarily have any access to a pasture? It is important to note that egg-laying facilities are allowed to label their products as "free-range" regardless of the actual amount of time the hens spend outdoors, nor is the range of outdoor space checked or regulated in any way. This label also doesn't guarantee that the hens had access to a natural pastured diet. Genuinely free-range hens and grass-fed cows are animals that roam freely outdoors on a healthy pasture where they can forage for their natural diet of seeds, wild plants, grass, herbs, and insects. These are now increasingly referred to as "pasture-raised" animals to avoid any prior confusion.

Additionally, it is important to buy these meats in their organic form because of the combined risks of pesticides, antibiotics, GMO feeds, and growth hormone exposure. You don't want your pasture-raised chicken to be eating processed foods!

When looking for the best, you want to buy pastured, grass-fed beef and free-range chicken that have been fed an organic diet. As an added bonus, given their freedom of movement, these animals' meat is naturally lower in saturated fat and higher in omega-3s as compared to that of their hormone-laden and often sickly conventionally raised counterparts. Obviously, the eggs and milk from these pastured animals are also safer. You've probably seen or eaten a farm-fresh egg, with a deep, bright orange color, and wondered why it looked so different from the typical supermarket egg? That's because the chicken that laid the egg was *healthy*! Same goes for milk.

Which brings us to cheese. When it comes to cheese, always check the label. Processed cheeses are a deal breaker. If you need a good reason to never eat sliced cheese singles again, consider that many of these "cheeses" aren't even technically made of cheese at all. Take a peek at the label and you'll find that they are actually "processed *cheese food*." Translation? Half of what you're eating is not cheese or food in the first place, but a harmful chemical concoction. Besides being made of poor-quality, contaminated milk, these dairy products are full of emulsifiers, refined vegetable oils, trans fats, and all sorts of additives like starches and gums that make up the other half of your "cheese." Even worse, when the label reads "processed cheese *products*," it means that they contain even *less* cheese, using various industrial powders and mixed solids instead.

It wasn't until I first moved to New York that I experienced processed cheese. As an Italian, I was surprised to find it on my pizza, no less! Friends took me to a local pizza place and ordered a large cheese pizza. When it arrived, it was covered with a white, chalky substance that I was told was "mozzarella." As someone familiar with what mozzarella is and isn't, I was stunned at the imposter that was passing for it on this side of the Atlantic. Although not all pizzerias serve processed cheese, this was one of my first experiences finding it on my plate, which led to a revelation.

I'd much rather eat organic full-fat cheese than any other available option—including low-fat. Surprised?

Once again, quality trumps quantity in my book. There are cheeses that are naturally low in fat and there are full-fat cheeses that get industrially stripped of a good part of their fat content. These low- or reduced-fat cheeses usually suffer from many of the same problems as processed cheeses. Since low-fat milk has very little flavor or nutrients, manufacturers add sugars, starches, and additives in an attempt to recuperate the lost texture and flavor of the cheese so that it tastes closer to that of its natural full-fat version. That goes the same for yogurt. Most commercial yogurt is chockful of artificial colors, flavors, additives, and high fructose corn syrup, which instead of delivering any health benefit at all, actually nourishes disease-causing bacteria, yeast, and fungi in your gut. Since your gut can only host a limited number of bacteria, eating these foods pushes the good guys out while letting the bad guys in. The inevitable result is that you get sick either initially or over time.

Did I just suggest that you eat full-fat cheese and yogurt? I sure did. Feel free to enjoy a cup of organic, plain, full-fat yogurt every day if you like. The best practice would be to make your own yogurt, or at least buy it fresh directly from the farm. Alternatively, some of my favorite commercial brands are Maple Hill Creamery (100 percent grass-fed variety), Stonyfield, Ronnybrook, Liberté, and Redwood Hill Farm. You can find them in many supermarkets and just as easily online.

A much better alternative to both high-fat and commercial low-fat dairy is to focus on those milks and cheeses that are naturally low in fat. For example, goat's milk and sheep's milk contain less fat than cow's milk while at the same time containing more protein and nutrients, allowing you to feel fuller longer. Dry cheeses like Pecorino or Feta are good examples. Next time you're about to eat a large piece of rubbery, processed mozzarella, try having a smaller piece of fresh, creamy goat cheese or a sharp Romano instead. These are but a few of the endlessly better choices.

RESTORE THE BALANCE OF PRO- AND ANTI-INFLAMMATORY FOODS

In addition to increasing our consumption of anti-inflammatory foods, we need to pay attention to which foods are producing inflammation in the first place. The typical Western diet is loaded with pro-inflammatory foods and these foods are actually speeding up brain aging. The most age-accelerating foods a person can eat tend to be those that are highly acidic—but maybe not in the way you might imagine. It's actually those foods loaded with refined sugars, refined grains, highly processed products such as margarine and soft spreads, and even alcohol that deliver the strongest acidic punch.

Though processed foods are convenient, more often than not convenience means large amounts of hidden trans fats, sodium, and nutritionally empty sugars, which are among the most dangerous ingredients you can ingest. Even seemingly harmless foods like "fortified" cereals and breads are actually masking overprocessed and mass-marketed products. While it is true that these foods contain some added vitamins and minerals, they are present only in their most synthetic forms, which means that their nutrients are destined to be poorly absorbed, if at all. Further, these products harbor excess hidden chemical fillers within their ingredients—including those preservatives and emulsifiers that are shown to damage heart, brain, and gut health alike.

Animal products like meat and high-fat dairy are also pro-inflammatory in nature, especially when fried in unhealthy oils. Nonetheless, if you limit your consumption to the appropriate portions and eat them sparingly and from safe sources, those foods are better for you than any processed food could ever be.

HELP YOUR DNA HELP YOU

Many nutrients have a strong effect on our genes. While some make us stronger, some weaken us instead. Fresh, organic produce, whole grains, wild-caught fish, and smaller amounts of pastured meats and eggs, along with natural sweeteners like fruit, honey, and maple syrup, are all ingredients to which our ancient DNA naturally responds positively. These same foods, especially the plant-based ones, have been linked to nothing less than brain rejuvenation, besides reduced inflammation, improved metabolic balance, improved insulin sensitivity, and a stronger immune system. On the other hand, pro-inflammatory, processed, and refined foods do quite the opposite, providing us with multiple side effects even at our DNA level.

These same foods also impact your microbiome. A diet that regularly includes prebiotic and probiotic foods will turn your microbiome into a champ, while one full of processed foods and fatty meat will achieve the opposite result. Too many people don't eat nearly enough fiber, nor do they think about consuming pre- and probiotics in their foods. The time has come to get a better feel for all the available options. Next time you go grocery shopping, and especially if you're suffering from digestive issues such as cramping, bloating, constipation, or diarrhea, bring the list in Box 2 with you for some healthy choices. Onions, asparagus, artichokes, and garlic are powerhouses when it comes to supplying the body with prebiotics. Additionally, be sure to stock up on fiber-rich carbs such as bran, oats, cruciferous vegetables (such as broccoli and cauliflower), berries, and all kinds of leafy greens.

Probiotic foods are less common than the prebiotic ones, but there are several options available to us that also happen to be really inexpensive. These foods contain "live bacteria," or probiotics, that upon reaching the intestine will replenish your microbiome and restore your intestinal health. As a rule, look for bitter and sour foods when it comes to feeding your gut's "good guys."

PREBIOTIC AND FIBROUS FOODS	OPTIONS OR INGREDIENTS
Vegetables	Broccoli, cauliflower, cabbage, root vegetables, artichokes, chicory, garlic, onions, leeks, asparagus, beetroot, fennel bulb, green beans, peas
Legumes	Chickpeas, lentils, red kidney beans, black beans, soybeans
Fruits	Bananas, berries, apples, nectarines, white peaches, persimmon, grapefruit, pomegranate, dried fruit (e.g., dates, figs)
Bread / cereals / snacks	Barley, rye, whole wheat, couscous, wheat bran, oats
Nuts and seeds	Cashews, almonds, pistachios, chia seeds, flaxseeds, psyllium seed husk

PROBIOTIC FOODS	OPTIONS OR INGREDIENTS
Fermented organic milk	Yogurt, kefir, and buttermilk
Pickled vegetables	There are many options, including sauerkraut, cabbage, turnips, eggplant, cucumbers, beets, onions, and carrots. Remember that the probiotic benefits are only present in unpasteurized foods pickled in brine, not vinegar.
Kimchi	A traditional Korean dish made of fermented cabbage, radishes, scallions, cucumbers, and several spices
Natto	A traditional Japanese dish made of fermented soy
Kombucha	A fermented drink made of sweetened black or green tea, along with friendly bacteria and yeast

Box 2. Prebiotic and probiotic foods

If none of these foods appeals to you, or if you consume them only on occasion, you need to take a probiotic supplement. Supplements come in capsule and liquid formulas and vegan versions are also available. But there's some misinformation out there that you need to be aware of. Many retailers would have you believe that the quality of a probiotic is based on just how many billions of bacteria are present, but in reality, it isn't the quantity as much as the *diversity* of bacteria that really makes the difference. Because different bacterial strains have slightly different functions and tend to colonize in different parts of the gastrointestinal system, probiotics that contain multiple strains are usually more effective. Check the labels, and make sure your supplement contains at least three of these major bacteria types: *Lactobacillum acidophilus*, *Lactobacillum helveticus*, *Lactobacillum rhamnosus*, *Bifidobacterium longum*, *Bifidobacterium bifidum*, and *Streptococcus thermophilus*.

Fortunately, once you change your diet, it only takes a short time for the microbiome to change as well. However, these positive changes can disappear just as quickly if a good diet is abandoned for a poor one, so make sure these foods and supplements become a regular part of your diet.

TRANSFORM YOUR FOODS INTO NUTRITIONAL POWERHOUSES

Following a diet that calls for whole or minimally processed foods is essential if we want to be able to access the nutritional content of those foods. It's just that simple. As you might recall, our world's centenarians *never* eat processed or packaged foods. All their food is fresh, which is a bit of an understatement, since it's often just picked from the garden or grove. This level of freshness ensures that their foods are as nutritionally dense as possible.

Food processing, on the other hand, is a major problem for our society. The milling of wheat, polishing of rice, filtering of cornmeal, and refining of cane sugar are all examples of methods that strip our food of most of its vitamin, mineral, and fiber content, since they are all contained in the outer layers and husks that are being discarded. B vitamins are especially affected by this process, with losses that can be as high as 50 percent. Freezing and canning are other processing methods that result in the loss of nutrients.

Modern farming is another big issue when it comes to the authenticity of our foods. The nutritional content of the soil in which vegetables and fruits are grown determines the food's final nutrient content. New kinds of crops have been scientifically developed to keep up with the demand for greater and greater yields, as well as to increase the food's resistance to pests and climate change. While these new varieties have produced bigger, faster-growing plants, their ability to take up a comparable quantity of nutrients has been lost in the process.

A landmark study on this topic looked at USDA nutritional data for over forty different vegetables and fruits from 1950 to 1999 and discovered significant declines in the amounts of vitamins (especially Bs and C), and minerals such as iron. These findings raised a huge debate that provoked subsequent contrasting reports. While some studies showed that the modern carrot does not contain even half the nutrients that would have been present in the carrots that our grandparents ate, others indicated that the vitamin content of these vegetables is broadly the same. So what's the truth of the matter? In my opinion, even if the latter studies detect similar vitamin levels, what they neglect to mention is that it is the *quality* of the reported vitamin content that really needs to be examined. Containing similar levels of one vitamin does not mean that the quality of the vitamin is the same—nor does it verify that the food's accompanying arsenal of phytonutrients is still intact and sound. Add to that the unrestricted use of pesticides and fertilizers on these crops and the result is the latest generation of a fast-and-furiously grown, pest-resistant pseudo produce. Full of preservatives and unnaturally perfect in appearance, this produce might beguile us, but unfortunately, it does not deliver.

Regardless of controversies, when their blood is tested, too many Americans turn up with nutritional deficiencies. After reviewing many of these tests, I know this to be a fact. There are no two ways about it. Those of us who are dedicated to taking care of ourselves and our families, and want to get the most nutritious produce possible for our tables, would do well to buy regularly from local organic farmers or health food stores.

The *way* you eat your foods also makes a difference. For the most part, vegetables, fruits, nuts, and seeds are at their most potent when eaten raw. In general, it is in this state that their enzymes, vitamins, minerals, and phytochemicals are still intact, making your produce as powerful as possible. Buy the freshest produce you can find, all the better if it's local and in season, and then eat it promptly! People worry a lot about the nutrient loss via the cooking process, but a vegetable can lose just as much nutrition languishing in the crisper as it does on the stovetop.

That said, there are vegetables that actually *prefer* the stovetop. Sometimes cooking increases the nutritional content of a food (especially of some vegetables) by breaking down the plant's cell walls and releasing the nutrients that would otherwise remain locked inside. Beta-carotene (the antioxidant responsible for the carrot's bright orange color) and lycopene (responsible for the tomato's ruby hue) are among these prisoners. Steaming, using a minimal amount of water, or lightly roasting your carrots and tomatoes will boost their levels of antioxidants, thereby providing your hungry neurons with extra protection against aging. Just be sure that you don't overcook the vegetables until they're mushy. Al dente is the name of the game for optimal nutrition.

While whole grains and legumes would be impossible to eat without some cooking, here's a great trick that will transform these foods into nutritional powerhouses: *sprout them*. The simple process of sprouting releases an army of healthy enzymes in seeds, legumes, and grains. This makes them easier to digest, allowing us to absorb their proteins, vitamins, and minerals more efficiently, while at the same time drastically reducing their cooking time. It's a win-win.

If you're new to sprouting, this is how you do it. Let's start with grains. They must be whole with the hull still on. Wheat berries, amaranth, barley, buckwheat, kamut, millet, rice, rye berries, sorghum, and spelt—all sprout beautifully. Spelt grains are a household favorite of ours. These grains contain the bran, germ, and endosperm of the spelt kernel, which translates into maximum nutritional payload. In order to sprout them, the first step is to soak them in water. This increases their moisture content while neutralizing their *phytic acid,* a substance that can cause bloating. You then drain the grains and keep them moist inside a Mason jar for a period of one to five days. Cover the jar with a piece of mesh or cheesecloth so that you can drain them easily, also allowing air to circulate to prevent molding. Soon you will see them sprout before your very eyes. Sprouted grains can be eaten raw or lightly cooked, and are even found in some healthy breads. The same process can be used to sprout legumes like lentils and mung beans and all sorts of seeds from sunflower to quinoa.

Let's move on to animal products. With the exception of extremely high-quality products such as sushi-quality fish, animal products must be cooked to kill bacteria and other pathogens. To bring out their full nutrient potential, fish and eggs are best consumed steamed or poached, with a dash of seasoning and fresh herbs. Meat is another food that cooking renders more easily digestible and bioavailable. For meat, roasting, grilling, or sautéing in a wok are ideal, while broiling and frying will increase their inflammatory and AGE-producing effects instead.

Regardless of which preparation method you choose, be stringent about selecting only healthy oils to use in your daily regimen. Most vegetable oils are available in both refined and unrefined (e.g., virgin) varieties. Much like grains and sugar, refined oils are made using highly intensive mechanical and chemical processes to extract the oil from seeds and nuts. This process removes nutrients, especially minerals, and creates a final product that is prone to oxidation. Oxidation makes refined oils more likely to produce free radicals in your body, which in turn work to age you. If this weren't bad enough, many refined vegetable oils are also hydrogenated. As we have discussed, the hydrogenation process transforms these oils into trans fats, making them thicker at room temperature so they can be sold as margarines and shortenings.

These oils might indeed be cheaper but will have a very negative impact on your health.

On the other hand, unrefined oils maintain the same qualities they possessed when still inside the plant. They have a full, rich flavor and are high in nutrients. Extra-virgin olive oil, for example, contains antioxidant polyphenols along with the healthy fats that would have otherwise been removed via the refining process. Whether you prefer olive or coconut oil, always choose the unrefined versions. Also keep in mind that oils rich in precious unsaturated fat such as olive oil and flaxseed oil are best consumed *a crudo* (raw, unheated), thereby retaining maximum nutrient content. However, if you use these oils for cooking, make sure you don't overheat or burn them (you don't want your oil to be sizzling and turning brown). For example, olive oil burns at 400°F (or 200°C).

For cooking, choose oils and fats with a high smoke point. The smoke point indicates the temperature limit up to which that cooking oil can be used, so the higher, the better. The best unrefined oil for cooking is avocado oil, which has the highest smoke point of all (520°F, or 270°C). Canola oil (rapeseed) and coconut oil come next. If you prefer animal fat for cooking, ghee (Indian clarified butter) is the best one (485°F, or 251°C).

EAT FOODS, NOT NUTRIENTS

Recent research, including my own work, supports the notion that nutritional supplementation is *not* equivalent to gleaning our nutrients from the whole foods we eat. In other words, supplements should be used to "supplement." This brings us back to the touchy subject of obtaining precious brain-nutrients from our foods versus our supplements. Too many people do not obtain sufficient vitamins and minerals from food alone. However, rather than improving their diets to reach their full nutritional requirements, they prefer to resort to supplements as a shortcut.

As we have seen in the previous chapters, there is increasing scientific evidence that supplements alone won't work. How our bodies respond to diet is a complex and highly coordinated system, and apparently what they can wrangle from a pill does not equal the benefits derived from real food. Think of nutrients as belonging to a team. A single player's skill, no matter their expertise, means nothing unless there is an equally adept communication and coordination with the other teammates. The same principle applies to food. Nutrients from food are superior to supplements because they act in synergy with one another and within our bodies in a way that supplements apparently can't. This downfall makes supplements the second choice when it comes to achieving our nutritional aims.

I'm not a big proponent of taking many supplements, as I believe the majority of our nutrients can and should come from the natural foods we eat. However, if you haven't been eating enough nutrient-dense foods, supplements can help correct and avoid possible deficiencies. It is important that you discuss whether you need to take supplements with your doctor. The following supplements can be useful if your diet is low in some (or all) of the brain-essential nutrients we discussed throughout the book:

- Omega-3 DHA: Supplementing with 300 to 500 mg/day of omega-3 DHA is a good idea, especially for people age sixty or older. If you don't eat fish, it is crucial to supplement your diet with at least 500 to 1000 mg of DHA per day.*

---

* Caution: Omega-3s can have a blood-thinning effect and might increase the action of blood-thinning medications such as warfarin and aspirin. Too much omega-3 can also result in bleeding and bruising. Always consult a health-care professional before adding omega-3 supplements to your diet.

- Choline: Most people in the United States don't consume enough choline-rich foods, especially fish. On those days you don't eat fish, eggs, or other choline-natural sources, consider supplementing with 300 to 600 mg/day of alpha-GPC (the most bioavailable form of choline) or 420 mg/day of *phosphatidylcholine* (a good source of both choline and omega-3s) as a memory booster.
- B vitamins: It is important to supplement your diet with a B complex that contains a full spectrum of B vitamins. B vitamins are well-known nerve tonics that help reduce stress and fatigue, and play a crucial role in the production of neurotransmitters. Additionally, if you are fifty or older, and/or suffer from gastritis, reduced stomach acid, Crohn's or celiac disease, or take medications like metformin (for diabetes) and acid blockers, talk to your doctor about having your vitamin B status checked. Especially as we get older, our metabolism naturally slows down and absorption of some vitamins like vitamin B12 might decrease as a result. Make sure it includes at least 50 mcg of vitamin B12 (cobalamin or methylcobalamin, as directed by your doctor).
- Minerals: Everybody needs to pay attention to their minerals, especially copper, iron, and zinc. Too few slow the brain down. Too many might rust your brain cells. Most people in the United States consume plenty of these minerals from everyday foods as well as other sources. If you are taking a multivitamin, check the label for its content of copper, iron, and zinc. If possible, choose a multivitamin without these minerals. Otherwise, check the label and make sure your supplements contain no more than 50 percent of the daily recommended dose (% DV) for these nutrients. Since dosages depend on your age and gender, refer to current Institute of Medicine's guidelines for specific values (http://www.nationalacademies.org/

hmd/Activities/Nutrition/SummaryDRIs/DRI-Tables.aspx) and discuss this with your doctor.

- Probiotics: Be sure to eat probiotic foods (e.g., yogurt, sauerkraut) every day. If you prefer to take supplements, or on days you don't have access to those foods, remember to select a high-quality probiotic supplement that includes a minimum of three bacteria types, as described earlier. Personally, when I don't have access to probiotic foods, I take my probiotic supplements in a refrigerated liquid form, as they are gentler on the GI tract.
- Antioxidants: If your typical diet hasn't been focused on fresh vegetables, fruits, nuts, and seeds, your brain might be running low on antioxidants. In order to replenish your antioxidant reserve, consider supplementing with 100 to 200 mg/day of coenzyme Q10 (CoQ10 or ubiquinone).* This is particularly important if you are sixty or older. CoQ10 is a powerful antioxidant involved in key metabolic reactions and energy production within brain cells.
- Herbs: If you lead a fairly sedentary lifestyle, ginkgo biloba, one of the world's most renowned brain tonics, could help boost mental clarity and sharpness by improving oxygen delivery to the brain. The recommended dose is 240 mg/day of ginkgo extract. Panax ginseng could also be helpful for memory support, especially if you tend to have high blood sugar levels and/or high cholesterol. This herb has celebrated anti-aging properties, and has been shown to lower both sugar and cholesterol levels. The recommended dose is 4 grams/day of red Panax ginseng powder. You can also find it in liquid form, mixed with honey and royal jelly.

---

* Caution: CoQ10 might interfere with blood-thinning medications like warfarin and aspirin. Speak with your physician.

WATCH YOUR COOKWARE

As previously discussed, some minerals like copper and aluminum can enter the body via unexpected sources such as water pipes but also your pots and pans. Personally, I don't use aluminum cookware at all. I also don't recommend use of any cooking plastics (for example, microwavable plastic containers) or synthetic nonstick surfaces like Teflon, which contains PTFE (*polytetrafluoroethylene*), a toxic compound. Instead, my kitchen is well equipped with stainless steel, glass, cast iron, and traditional ceramic. Needless to say, disposable plates and paper cups should be altogether eliminated or at least used as a last resort.

13

A Typical Brain-Healthy Week

DIETARY GUIDELINES FOR OPTIMAL COGNITIVE FITNESS

Ready to put all this information into practice?

Let's take a look at the Brain Diet Pyramid in Figure 2 on the next page. This chart is a snapshot of the types and proportion of foods we should eat on a regular basis for optimal brain health. Along with the core brain-food groups, it contains healthy beverages, fats, and natural sweeteners, arranged according to how much they contribute to a neuro-nutritionally balanced diet.

On a weekly basis, make sure you eat plenty of those foods indicated at the base of the pyramid. The numbers along the margins will remind you of how many times a week you can have these foods. To start, drink at least eight 8-ounce glasses of filtered or genuine spring water and/or herbal tea every day. At the same time, make sure you eat leafy green or cruciferous vegetables every day. When it comes to these vegetables,

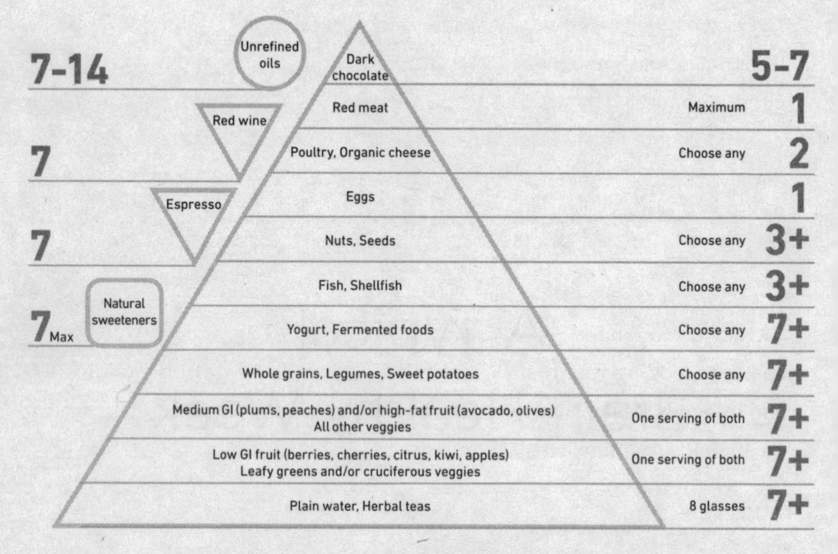

Figure 2. The Brain Food Pyramid: recommended foods and number of servings per week

the more, the better, but certainly no less than 1 cup a day. Add an additional vegetable serving a day such as carrots or onions, and remember to include high-fat fruits like avocado and olives on a regular basis. Typical serving sizes are one-quarter of an avocado and 4–5 olives.

Enjoy a piece of low GI fruit, like berries, cherries, oranges, grapefruit, apples, and pears, every day. If you prefer medium GI fruits like plums and peaches, that's fine, but keep an eye on portion size: 1 cup of low GI fruits versus ½ cup of medium GI fruits. (Refer to chapter 6 for a complete list.)

Aim to eat up to two servings of whole grains and legumes at least once a day. And don't forget your sweet potatoes. To get a visual of serving size, that's 1 cup of cooked grains or legumes, or 1 slice of whole-grain bread, or one-half of a small sweet potato (the size of your fist). See how easy it is to eat two servings of either?

Organic plain yogurt and fermented vegetables like sauerkraut or brined pickles should also be consumed on a daily basis, about ½ cup of either (or more).

Wild-caught (or at least antibiotic-free) fish is a brain must. Shoot for one serving (2–3 ounces) three times a week and focus on fatty fish like salmon, mackerel, trout, herring, Bluefin tuna, sardines, anchovies, and striped bass. Fish eggs (1 tablespoon) and shellfish (3 ounces excluding the shell) are also great options.

Raw unsalted and unsweetened nuts and seeds are crucial when it comes to feeding your brain with healthy fats and hard-to-find vitamins and minerals. Make sure to eat three or more servings a week, focusing on almonds, walnuts, chia seeds, flaxseeds, and sunflower seeds. Serving size: 2 tablespoons.

Now for animal products. Eggs are great in moderation (1–2 eggs per week), followed by organic pasture-raised chicken (3 ounces per week) and organic cheese—preferably dry cheese like Pecorino or fresh goat cheese like Feta (1–2 ounces per week). Red meat (beef, pork) is not on our menu and should be consumed no more than once a month or on special occasions.

Herbs and spices should be used in preference to salt and in combination with healthy, organic, unrefined oils. Extra-virgin olive oil is a must. Flax, hemp, avocado, and coconut oils are also great choices. The recommended dose is 1 tablespoon, twice a day. Feel free to use them along with healthy condiments such as apple cider, rice, or balsamic vinegar; tamari (a naturally gluten-free version of soy sauce); coconut aminos; brewer's or nutritional yeast; and miso paste to further enhance your foods. Be generous with these condiments; they are good for you.

If you want to have an alcoholic drink, consumption should focus on red wine, preferably organic. If you're a woman, one 5-ounce glass of red wine a day is the way to go. If you're a man, you can have up to two glasses a day.

Coffee can be supportive of brain health for some people. My personal preference is 1 cup of espresso a day to provide your brain with the optimal balance of caffeine to antioxidants, but regular organic coffee works, too. If you take your coffee black, or with a dash of (organic) milk, that's excellent. If you need a little sugar, I'd recommend no more than 1 teaspoon of coconut sugar or other organic sweetener like stevia per cup of coffee. Although there is no clear association between drinking tea and a reduced risk of dementia, green tea has almost as many antioxidants as coffee, without the jitters—so feel free to have up to 2 cups a day.

Finally, I strongly recommend a small piece of dark chocolate just about every day (approximately 1 ounce). Dark chocolate is rich in antioxidant flavonoids, magnesium, potassium, and theobromine, the substance that supports blood circulation and makes you happy. Remember to focus on high-quality dark chocolate with at least 65 percent cocoa and little or no added sugar.

At the same time that you're increasing consumption of brain-healthy foods, make an effort to limit or, better, completely eliminate consumption of foods that are not so healthy or even downright harmful to your body and brain. These include:

- All fast foods
- White sugar, artificial sweeteners, and table salt
- Meats: red meat (beef, pork), bacon, deli meats, cold cuts, cured meat, bologna, salami, pastrami, and any other processed meat
- Processed cheese/dairy products such as canned cheese, American cheese, blue cheese (unless organic), sweetened yogurt, commercial ice cream and custards, string cheese, spreads, margarine, sweetened or flavored milk beverages, and any other processed cheese/dairy products
- Refined grains such as white rice, white bread, corn bread, cornflakes, breakfast cereal, commercial doughnuts, cookies, crackers, muffins, pies, cakes, and any other processed grain products

- Processed nuts including salted, honey roasted, or sweetened
- Condiments such as ketchup, mayonnaise, Worcestershire sauce, barbecue sauce, salad dressings, commercial spreads, soy sauce, and all refined oils (especially safflower oil, sesame oil, sunflower oil)
- Soda, energy drinks, fruit juice (unless fresh-pressed, in which case limit to 1 glass a day)
- Spirits like beer and distilled liquors. Limit consumption to 1 can of beer (12 fluid ounces) or 1 finger of liquor (hold your finger next to the glass and pour yourself only a finger's width), no more than one to two times a month.

If this sounds easier said than done, Table 9 illustrates what a typical weekly menu should look like.

For example, breakfast on Monday could be a hearty bowl of oatmeal topped with a handful of chopped walnuts, fresh blueberries, and maple syrup. Lunch introduces my long-time favorite, Ayurvedic Mung Bean Soup. A cup of yogurt and a handful of almonds make for a great afternoon snack. To round it all off perfectly, a nice piece of steamed Alaskan salmon served with homemade broccoli puree is an excellent way to end your first brain-healthy day. On day two, you'll start your day with organic goat milk yogurt topped with fresh blackberries (and a little honey if you like), and move on to a flavorful Brown Rice Mushroom Risotto that combines the distinctive aroma of wild mushrooms with the cheesy flavor of nutritional yeast for an extra vitamin kick. A little salmon roe over rice crackers makes for a yummy, omega-3-loaded snack, while my Grilled Sweet Potatoes with Fresh Spinach Salad will satisfy any sugar craving and also keep you full all night. And so forth. Does this sound like dieting? I didn't think so.

That said, feel free to mix 'n' match if you like. Swap some of the suggested meals with any of those included in chapter 16 or on my website (www.lisamosconi.com), should they seem more appetizing or interesting to you. Just be sure to exchange each food type for the same food type, like fish for fish or eggs for eggs, so as to preserve the nutrient balance over the course of the week—though you are welcome and encouraged to swap meat and poultry for fish any time you wish. More specific recommendations are provided in Step 3: Toward the Optimal Brain Diet.

	Breakfast	**Lunch**	**Snack**	**Dinner**
Monday	Cup of coffee or herbal tea Cereal and fruit (*Oatmeal with maple syrup, walnuts, and blueberries*)	Legumes and vegetables (*Ayurvedic Mung Bean Soup*)	Yogurt and/or nuts (*tamari-roasted almonds*)	Fish and vegetables (*steamed Alaskan salmon and broccoli puree*)
Tuesday	Cup of coffee or herbal tea Yogurt and fresh fruit (*Goat milk yogurt with blackberries*)	Grains and vegetables (*Brown rice mushroom risotto*)	Salmon roe over rice crackers	Sweet potato and vegetables (*Grilled Sweet Potatoes with Fresh Spinach Salad*)
Wednesday	Cup of coffee or herbal tea Cereal and fruit (*Buckwheat porridge with dried fruit and almond milk*)	Legumes and vegetables (*Chickpeas Tikka Masala with basmati brown rice*)	Yogurt and fresh fruit (*Italian plum*)	Fish and vegetables (*Miso sea bass fillet and dandelion greens*)
Thursday	Cup of coffee or herbal tea Yogurt and fresh fruit (*Goat milk yogurt with raspberries*)	Grains and vegetables (*Farro with pesto and zucchini*)	Spelt crisp bread with avocado	Poultry and vegetables (*Dad's Lemon-Roasted Chicken and roasted Brussels sprouts*)

	Breakfast	Lunch	Snack	Dinner
Friday	Cup of coffee or herbal tea Cereal and fruit (*Overnight oats with raisins, flaxseeds, and honey*)	Legumes and vegetables (*Lentil burger and green salad*)	Yogurt and/or nuts (*Brazil nuts*)	Cheese and vegetables (*Mediterranean chopped salad with Feta*)
Saturday	Cup of coffee or herbal tea Eggs (*Sicilian Scrambled Eggs*)	Grains and vegetables (*Rainbow Buddha Bowl with Maple Tahini Dressing*)	Apple and nut butter (*Organic almond butter*)	Fish and vegetables (*Fresh tuna salad*)
Sunday	Cup of coffee or herbal tea Cereal and fruit (*Avocado Toast*)	Grains and vegetables (*Minestrone with whole-wheat pasta*)	Sunday treat (*Mango chia pudding*)	Eggs and vegetables (*Vegetable omelette and watercress salad*)

Table 9. Sample weekly menu plan. All recipes can be found in chapter 16 and on my website at www.lisamosconi.com

DON'T SKIP BREAKFAST. DON'T SKIP MEALS. SKIP SNACKING INSTEAD.

Now for a few practical things to consider as you plan your weekly menu. Let's start with breakfast.

For the centenarians, it's a no-brainer. Breakfast is the most important meal of the day. Instead, in our society there is some confusion over whether breakfast is an important meal at all. Far from having what might be the centenarian's luxury of enjoying a good breakfast, most of us find the act of getting ready for the world each morning a blurry-eyed, caffeine-fueled affair. On top of that, many people rush out the door before having had the chance to put something in their bellies first.

I've encountered some interesting breakfast trends in the United

States. Quite a few people skip breakfast altogether in favor of a cup of coffee. Often, this is actually a "coffee drink" and comes with copious amounts of added sweeteners, artificial flavorings, and cream substitutes. On the other hand, just as many people like to wolf down large portions of refined white bread, cereal, or pastries, paired with commercial orange juice that is akin to liquid sugar. Needless to say, I don't recommend that either.

A good breakfast is exactly what your brain needs after a long night without food to literally "break (its) fast." By a "good breakfast," I don't mean a gigantic, heavy meal. What our brains need first thing in the morning is light, sustained energy. This can be achieved by way of glucose-rich foods that include a good balance of fiber, vitamins, and minerals, along with adequate protein and fat, and of course filtered water or tea to rehydrate.

If your favorite breakfast is cereal and orange juice, that's okay. These foods have been breakfast staples for generations in the United States, so it's only natural that you might enjoy them in the morning. However, let's be sure the focus is on quality. Juice is better made with one to two freshly squeezed oranges, which isn't too time-consuming once you get the hang of it. Cereal should be made of whole grains *without* added sugar. Read the labels. If a serving of cereal contains more than 5 grams of sugar, that's not good for you. Opt for unsweetened cereal instead and add your own honey, maple syrup, or coconut sugar, if need be.

Some people prefer something sweet to start the day, while others prefer savory foods. You'll find several brain-healthy options in the next chapter. Feel free to choose whichever works best for you, and at the same time, try some new things, too. You never know, you might find that some unexpected alternatives make you feel sharper and more focused than old habits.

It turns out many people skip other meals in addition to breakfast, especially lunch, which makes one wonder how they're managing to function at all. Part of the problem might be that their meals are not as appetizing or energizing as they could be. According to recent surveys, during workdays, Americans will commonly eat a sandwich consisting of any number of fillings between two slices of bread. The bread is typically white, and most fillings are deli meat such as ham and roast beef, cheese, tuna salad doused in commercial mayonnaise, or the classic peanut butter and jelly. Potato chips and soda are also commonly found in lunch boxes. Other common options include cafeteria food or fast food. Except for the lonely slice of tomato folded into the sandwich, are you seeing any brain foods in there? If you are into sandwiches, keep in mind that there's a big difference between processed ham, Swiss, and mayo on white bread—and grilled eggplant with spinach and hummus on whole wheat. As you will have guessed, I'd recommend the latter.

American dinners are generally more nutritious, though most often feature meat as the main dish with smaller side dishes of rice, potatoes, or pasta, and maybe a vegetable or salad. Chinese takeout, pizza, and frozen microwavable entrees turn out to be very popular choices. Many people also drink juice, soda, or beer and sometimes wine with their evening meal, although wine is still less common in the States than in Mediterranean countries.

The low nutritional quality of these meals might very well be what leads Americans to snack throughout the day. By and large, snacks consist of processed items such as candy, cookies, chocolate, crackers, potato chips, and pretzels. And let's not forget the sugary cocktails served at happy hour.

Overall, the typical Western diet is the opposite of "brain healthy." Time to roll up our sleeves and address this issue once and for all.

TAKEOUT. JUST SAY NO.

America is rapidly becoming a delivery nation. The long list of quick-service giants that offer delivery, from Starbucks to Dunkin' Donuts, not to mention the growing list of delivery-only restaurants, underscores an increasingly apparent truth about what Americans look for in their food: convenience. Families are working more than ever, which leaves very little time to spend in the kitchen. If this is you, beware. As fast and easy a solution as takeout seems, make it a habit and you will just as quickly pack on the pounds and actually harm your brain. By and large, takeout food is bad news for your brain (and your body, too, to be fair). Except for those restaurants that pride themselves on using solely high-quality organic ingredients, most takeout options include rock-bottom produce and commercially fed and industrially bred meat and fish. All of this adds up to your ingesting more hormones, antibiotics, and pesticides than genuine food and its nutrients. If that weren't bad enough, these foods are also typically cooked with refined oils and sugars, trans fats, and more salt and sodium than you can imagine. Even salad bars can be full of empty calories, artificial dressings, and stale preservative-laden products. That's not counting the additional commercial drinks and processed breads, fries, and condiments, or the fortune cookies and GMO soy sauces thrown in for free.

It is nothing less than crucial to the health of your brain that you don't make takeout a regular everyday habit.

I guess it goes without saying, then, that I hardly ever order in. When I do, I choose healthy restaurants that serve high-quality food and I am willing to pay extra for this once-in-a-blue-moon treat. To this day, on the rare occasion my family and I get takeout, it's sushi from a nearby Yelp-rated five-star restaurant.

That said, many people don't have access to healthy prepared-food options. I experienced this food desolation when I first moved out of Manhattan to relocate to Brooklyn. My new neighborhood was only a few subway stops away from the city, and although it felt similar in many ways, the option of dashing into a supermarket that contained a healthy prepared-food selection on my way home had vanished. With no higher-quality supermarkets in sight, I tried countless online prepared-food services but couldn't find one that was worth the money. If it's hard for me to find good options here in the New York area, I can't even imagine how difficult it must be in other parts of the country.

In the end, preparing our own meals is the best (if not the only) way to go. By cooking at home, you can ensure that you and your family eat fresh, safe, wholesome meals. You'll look and feel healthier, have more energy, and be able to maximize your brainpower and longevity. Contrary to popular belief, creating healthy meals doesn't have to involve a huge investment of time and effort. Here are a few practical tips:

1. Plan your meals ahead. Gather your recipes and make a list of all the ingredients you need *for that week*.
2. In the summer, favor salads and other raw foods that require minimal preparation. Avocado toast, hummus with crudités, a tomato salad with fresh basil, a cold beet soup, or smoked salmon (and caviar) tartines are all great options. Throwing some fresh vegetables and fish on the grill goes quickly once you're in the habit. Did you know that summer fruit grills and caramelizes beautifully? Peaches, pineapple, plums, even watermelon—they all know how to take the heat.
3. When you cook, make meals in bulk and freeze leftovers into single portions to defrost and heat up on those days you don't have the time or energy to cook.
4. Loading a slow cooker with vegetables, grains, legumes, fish, or chicken in the morning allows you to come home to a piping-hot meal at night—with minimal preparation or cleanup.
5. Cook once, eat for a week! By cooking enough veggies, grains, and protein once a week, you can use these foods to create quick and easy meals throughout the week.

Personally, I consider Sunday morning to be my "kitchen retreat" time. I have perfected my technique to the point that within a couple of hours, I manage to make a pot of brown rice, steam a large piece of fish, roast several pounds of vegetables, *and* make a soup. How? Here goes: I soak the rice (or any other grain) the night before, which not only brings out the nutrients but also makes the grains tender, cutting cooking time in half. While the rice is boiling, I turn on the oven and start chopping up the veggies. Brussels sprouts are a big favorite of my family's. All you need to do is wash and cut them, toss with avocado oil, and roast in the oven for twenty minutes at 350°F. Sometimes I add sweet potatoes to the mix, using coconut oil to bring out their flavor. While the rice is boiling and the veggies are roasting, I make soup. Most likely I'll make a root vegetable soup with carrots, parsnips, rutabaga, and butternut squash. In just five quick steps: (1) wash, peel, and cut the veggies (or buy them already cut, which in this case might be worth it, as it saves a lot of time), along with an onion and 2–3 cloves of garlic, (2) sauté in 2 tablespoons of coconut oil, (3) cover with vegetable broth, (4) cook over medium heat for about twenty minutes, and (5) puree with an immersion blender. It takes no time to make this delicious soup, which I serve with a drizzle of avocado oil and roasted pumpkin seeds, or with a dollop of coconut cream and cilantro (see my blog for the full recipe). By then the rice is ready, the veggies are roasted, and the soup is simmering in its pot. Steaming a few salmon fillets takes no longer than five to ten minutes. Whisking together a tasty tamari-scallion sauce for the fish takes just a minute more and we're ready to roll.

By the time my husband and two-year-old daughter are back from their morning stroll, I have enough food for up to five meals. And not just any food, but high-quality, fresh, vibrant, nutrient-rich food completely devoid of additives, artificial sweeteners, unhealthy fats, and sodium.

In the next chapters you'll find many more examples of what a typical brain-healthy week could—and should—look like from a nutritional perspective. The good news is all the recipes provided in this book are super quick and easy to follow. In fact, one of the keys to eating right for your brain is to make sure your foods are kept as close to their natural state as possible, which comes down to less preparation time, optimized nutritional quality, and more flavor.

GIVE YOUR TUMMY A BREAK

New research indicates that reducing the overall caloric content of the diet boosts cognitive capacity, reduces cellular aging, and promotes longevity. Additionally, as most Americans tend to eat too much to start with, paying more attention to portion size and mindless snacking can only help. One of the best ways to eat less and maintain nutritional quality is of course to increase your consumption of low GI fruit, vitamin-rich vegetables, and lean protein, while reducing that of sugary and fattening foods.

As we discussed in chapter 9, intermittent fasting can provide many important health benefits, from "rebooting" your metabolism to helping your body burn fat more effectively. Don't let the word "fasting" intimidate you—it's much easier than it sounds. Of the many intermittent fasting schedules available, overnight fasting is my personal favorite. You simply give yourself a ten-to-twelve-hour break between dinner and breakfast, during which you are not eating or snacking. By rights, you should be sleeping, or at least resting. This simple practice has been shown to reduce adipose body fat, improve insulin sensitivity, and protect against obesity and diabetes, which are known risk factors for cognitive aging and dementia.

That said, this practice might not work for everyone. Take the test in chapter 14 to find out whether or not you're ready for this practice. I generally recommend specific forms of intermittent fasting to those who score at the Intermediate or Advanced levels but not necessarily at the Beginner level (though they are of course welcome to give it a try). In any case, be sure to talk it over with your doctor.

AND WHEN YOU'RE DONE EATING . . .

Walk Away from That Couch

Regular physical activity is essential to keep your brain healthy. Additionally, some studies have shown that people who engage in regular physical activity have a healthier microbiome, reduced inflammation, and higher levels of brain-protecting hormones than sedentary people.

Research to date shows that we have two main options when it comes to boosting the neuroprotective effects of exercise. The first option is to engage in vigorous activities for at least one hour a week. Exercise is considered vigorous when your heart beats fast enough that talking normally becomes difficult. If you were asked to sing at the same time, you would probably blow the audition. These activities include things like jogging, playing tennis, handball, swimming, hiking, aerobic dancing, or even lifting weights as long as it's mixed with more movement. Anything that makes you sweat and run out of breath will do.

Not your thing? No problem. Moderate-intensity activities will work as well—as long as you exercise a bit longer and more often. Moderate intensity means that your heart will beat faster than normal. At this raised rate, you can carry on a conversation (albeit sounding a little breathy), but you probably can't belt out a tune. Some great options include power walking, bicycling, gentle swimming, or even doing household chores and gardening, as long as they make your heart beat faster. Shoot for thirty to forty-five minutes five times a week, or a total of two to three hours a week.

Choose the option that works for you, and make it an integral part of your lifestyle. More specific recommendations await you in the next chapters.

Keep That Brain of Yours Busy

It is also useful to exercise your brain in simple ways that strengthen the connection between brain cells and counteract the brain shrinkage that would otherwise occur naturally as we age. There are plenty of options, from reading this book to going to the theater.

My personal suggestion is to try board games. As we discussed in chapter 10, playing board games is an excellent way to spend time with your family and friends while engaging intellectually at the same time. In Italy, I grew up playing Tombola (or, as you probably call it, Bingo), with my grandparents and friends alike, which always made for a fun, cozy Sunday afternoon or a memorable Christmas Eve. Now that I'm older, I play Tombola with my nieces and *their* grandparents, and can't wait for my daughter to be old enough to join in. Whichever form of entertainment you choose, play hard.

Also remember to make time to see your friends and family. Looking for inspiration? Go to the movies or to a play. Go for a walk if the weather is nice. Check out a new vegetarian restaurant. Cuddle on the couch and peruse old family pictures, or just talk about your day. Call your family and friends if you can't get together in person. Anybody you haven't seen in a while? This is a great time to say hello and catch up. Having your support team around you on a regular basis is essential for making lifelong memories and making you (and your brain) feel that all is well.

Sleep Tight (Don't Be Afraid to Take a Nap)

Make sure you sleep well, especially during the first part of the night (relative to how long you can sleep) when your deep sleep stage occurs. This is essential for the brain to cleanse itself from harmful toxins, including those amyloid plaques characteristic of Alzheimer's.

If you aren't getting enough sleep, consider taking naps during the day to get some rest. For years, napping has been derided as a sign of laziness. In America, people are "caught" napping or "found asleep at the switch." In many cultures, particularly those that hail from Mediterranean and tropical regions, afternoon napping is commonplace and is actually built into the daily routine. When I was little, my *nonno* (grandfather) would always excuse himself to take a catnap right after lunch. Siesta is a big deal in many other countries the world over. Although nap time is not officially scheduled in these countries, it is not uncommon for stores and government offices to close an hour or two every afternoon so that people can take a break and relax (especially during the hot summer months). While this might or might not negatively impact productivity, napping has recently garnered a new respect in the United States thanks to scientific evidence that dozing off benefits both mental acuity and overall health. Give it a try and see if it works for you.

Stop and Smell the Roses

Finally, if you are not feeling happy or satisfied with the way your life is going, you can eat all the broccoli you want and your health might suffer anyway. It is very important that you take whatever steps necessary to reduce stress and enjoy your journey on the planet as best you can.

Take a breather, enjoy time with your family and loved ones, and make a point of being social in those ways that feel best to you. Seek out those communities that make you feel needed, appreciated, and productive. Finding your happy place is a must when it comes to your overall health *and* the well-being of your brain.

STEP 3

TOWARD THE OPTIMAL BRAIN DIET

14

How Brain-Nutritious Is Your Diet, Really?

THE THREE LEVELS OF NEURO-NUTRITION CARE

Each brain is unique and requires different degrees of care. The best way to determine what your brain needs for optimal nutrition and health would be to conduct several thorough exams, not the least of which would be a brain scan. However, not many people have access to a brain scan to take a peek at what's actually going on inside their heads. This prompted me to develop a test to help predict what your brain might look like if you had that personalized scan before you.

The test is based on my experience inspecting hundreds of brain scans over many years of research. This familiarity has allowed me to observe certain correlations between what shows up in the scan and a variety of clinical and lifestyle measures, especially diet and nutritional quality.

While the test won't illustrate what your brain actually looks like, it will help pinpoint how this most precious organ is feeling *on the inside.* For example, some brains have been fed pro-inflammatory foods for so long that they are now in dire need of soothing that inflammation. Other brains receive so few omega-3s that their neurons can no longer function properly. Others still grow desperate for oxygen because of a sedentary lifestyle having deprived them too long of what they need to comfortably survive.

If you think back on chapter 1, you might remember the difference between the brain scans of one person who had followed a brain-healthy diet and another person who did not eat well at all. You probably recall that the brain of the first one looked great—while the other brain showed signs of aging and deterioration. The simple truth is that if you feed yourself high-quality nutrients, your brain in turn will be made of high-quality tissue. This is what helps you stay cognitively fit and provides you with resilience against aging and disease.

The test will prompt you to take a careful look into various choices you make on a daily basis to get a clearer picture as to which ways you support your brain health and which ways you might inadvertently sabotage it instead. As I've made clear throughout the book, eating well for our brains is difficult because the clues don't arrive in the form of jeans that fit or firmer abs. So taking the test is a convenient way to learn how close or how far away your daily diet is from one that's optimal for long-term cognitive fitness.

As a result, the test will classify you at one of three main levels: Beginner, Intermediate, or Advanced.

Your current level is a reflection of all the choices you've made and continue to make on a daily basis with respect to your diet, nutrition, and overall health. This information will in turn help you zone in on which aspects of taking care of your brain health need work and which ones are already pointed in the right direction—allowing you access to a health plan that is specific and detailed to your individual needs.

When it comes to getting healthy, one size does not fit all. Having as much information as possible with regard to your own brain's health status can be extremely helpful in guiding you to take powerful steps toward gaining balance and achieving your goals.

So along with the general recommendations described in Step 2: Eating for Cognitive Power, once you discover which level you're at, the next chapter will provide you with targeted recommendations that are customized to your level in particular. These recommendations are aimed at providing natural lifestyle guidelines that can be followed at home and on a day-to-day basis. In addition, there are brain-healthy recipes and sample menu plans that will further help you take control of your own customized brain health. This final step of discovering and embracing your personal care will give you all the necessary information to achieve and maximize optimal brain power.

TEST INSTRUCTIONS

The test is focused on your diet, although it also gathers some information about your behavior, beginning with those habits you had as a child and continuing to the basic patterns you've established since. Base your answers on what you've observed to be the most consistent indicators of your behavior over a long period of time, rather than what might be new or recent to your present state. If you developed an illness in childhood or as an adult, think of how things were for you before or after that illness, focusing on what is more reflective of what your habits are overall. For fairly objective physical traits, answering clearly will be less of an issue. Since behaviors tend to be more subjective, answer according to how you have acted most of your life or predominantly during the past few years.

For each question, circle the one response (A, B, C, or D) that best applies to you.

In some cases, you might find that none of the responses offered describe you exactly. In those instances, don't worry. Just choose the answer that comes closest to describing your general tendencies. Remember, we're looking for common patterns and habits, so there's no need to get hung up on the exact details or specific wording of each question or response.

If for any given question you feel that more than one of the responses applies to you, circle the one that applies the best. All the words in that column need not apply for you to make the selection. For example, in response to the question "What are you most likely to have for breakfast if time is no object?" one of the answers is: "Something along these lines: eggs, bacon, or sausage, pancakes, buttered toast, hash browns." As long as one or more of the foods listed applies, select that response.

There might be questions that are difficult for you to answer. For example, if you have a peanut allergy, you won't be able to pick your favorite peanut butter type. In this case, just try to imagine what you *probably would have* picked if you weren't allergic. Would you buy one of the leading commercial brands or a certified organic brand? Would you eat a spoonful as a quick snack, or happily work on the entire jar?

Similarly, vegan/vegetarians might be hard-pressed to answer some questions about meat consumption. In this case, just try to imagine what you *would* do if you *were* to eat meat instead. For example, if you're the type who regularly eats organic produce, you might very well also be the type who would have chosen organic grass-fed beef had you been a meat eater.

Also, be sure to answer all questions in terms of what you *actually* do, not what you would *like* to do or think you *should* do. Try to be as thoughtful as you can, and remember that this is not about being right or wrong or good or bad. These are your private answers, so the only "right" way to answer these questions is with honesty. You won't get grades or a lecture. We are here to define your typical lifestyle patterns and choices as they stand now to improve whatever can be improved for you to enjoy your future good health.

You might be surprised to find out that you don't know the answers to all the questions. For example, you might not know offhand how you would react to a specific type of food or combination of foods. If this is the case, what you should do is simply put the test aside for a little while until you can gauge your reaction to the food in question. Though I don't want you to struggle too much with any questions or aspects of this test, accuracy is important, so take your time and don't rush through them.

Note that you can always take the test again at any point in the future. In fact, this is something you'll want to do periodically, to see if your habits have changed, which can *and will* occur as you improve your choices to better suit your brain's needs.

TAKE THE TEST

		A	B	C	D
1	**How often do you eat fish?**	Rarely or never. I don't like fish.	A couple of times a month.	At least once a week.	I love fish. I eat it on average twice a week, if not more.
2	**How often do you eat salmon, sardines, mackerel, or fresh tuna?**	Hardly ever or never.	I'll have salmon a few times a month.	Often, like once or twice a week.	More than twice a week.
3	**When you eat fish, how is it usually prepared?**	Canned tuna fish or a tuna salad sandwich is okay.	Fish with some kind of sauce, like sweet and sour or teriyaki, for example.	I'll often have sushi or something like that.	Roasted, steamed, carpaccio, any which way!
4	**How often do you eat caviar or fish roe?**	Ewwww.	Once in a blue moon.	A couple of times a month.	About once a week, or as often as I can.

PAGE TALLIES

A=	B=	C=	D=

		A	B	C	D
5	**Do you take fish oil or omega-3 supplements?**	No, I don't.	Sometimes (or as prescribed by my doctor).	Frequently, like two to three times a week (or I take them when I don't eat fish).	Yes, every day (or I take them when I don't eat fish).
6	**What are your favorite vege-tables?**	Potatoes! Tomatoes are fine, too, especially in pizza or pasta sauce.	Anything goes, as long as there's a tasty-enough dressing mixed in or on top.	Broccoli, green beans, and/or Brussels sprouts.	Fresh, vibrant summer salads with all kinds of veggies thrown in.
7	**How often do you eat orange vegetables like carrots and sweet potatoes?**	Rarely. I'll sometimes order sweet potato fries instead of regular fries.	Sometimes, mostly in soups.	About once a week, in my salads or as a side dish.	A couple of times a week. Roasted, mashed, or steamed, they make a great side dish.
8	**How often do you eat leafy green (broccoli, kale, spinach, etc.) and/or cruciferous vegetables (broccoli, cauliflower, or cabbage)?**	Rarely. I don't eat a ton of greens, but I eat the classic iceberg lettuce salad some-times.	I enjoy them once in a while, but I like other veggies better.	Frequently, like two to three times a week.	Every day or almost every day.
9	**How often do you eat low GI fruit like berries (blueberries, raspberries) and/or citrus (orange, grape-fruit)?**	Rarely or never.	Once in a while—not regularly.	Once or twice a week.	Twice a week or more.
10	**What kinds of fruit do you prefer?**	Bananas, figs, raisins, and/or dried cranberries.	Tropical fruits like mango and pineapple.	Apples, pears, nectarines, cantaloupe, and/or watermelon.	Berries and/or citrus fruits like oranges.

PAGE TALLIES

A=	B=	C=	D=

		A	B	C	D
11	**When you go shopping for produce, what are you most likely to buy?**	I usually buy frozen fruits and vegetables.	I typically buy lettuce and tomatoes or other salad ingredients.	Leafy greens like kale and/or cruciferous vegetables like broccoli.	All sorts of different types of colorful veggies—a rainbow on a plate!
12	**How often do you eat frozen fruits or vegetables?**	Often, like three to four times a week or more.	Frequently, like once a week.	Occasionally, like once or twice a month.	Rarely or never.
13	**Do you make an effort to eat fresh, organic produce every day?**	Not really.	I'm not overly concerned with organic foods, but I'll eat them when available.	Yes, as much as I can afford.	Definitely!
14	**How often do you eat fatty fruits like avocados and/or olives?**	Rarely or never.	Sometimes, like a few times a month.	Frequently, like once a week or so.	Once a week or more.
15	**How often do you eat whole grains (e.g., brown rice, quinoa, bran, wheat)?**	Rarely. I prefer regular white-flour foods.	Sometimes.	Often, like three to four times a week.	Almost every day.
16	**How often do you eat white-flour breads, pastas, or pizza?**	Every day or close enough.	About three to four times a week.	I like my pasta, but everything in moderation.	Rarely or never.
17	**How often do you eat legumes like lentils, beans, or chickpeas?**	Rarely or never.	A couple of times a month.	About once a week.	Twice a week or more.
18	**Are nuts, especially almonds and walnuts, a regular part of your diet?**	Not really—unless you count peanuts. (In that case, yes!)	Sort of. I like to bake with them.	I often put them in my shakes.	I eat them all the time, as a snack and in my soups and salads.

PAGE TALLIES

A=	B=	C=	D=

		A	B	C	D
19	**Are seeds, especially flax, chia, sesame, or hemp seeds, a regular part of your diet?**	I barely recognize what some of those are.	I eat them occasionally.	I often put them in my shakes and/or salads.	Yes, I love to include them in my meals.
20	**How often do you eat plain, unsweetened yogurt?**	Oh dear. Yuck.	I don't love plain yogurt but will eat it with fresh fruit or other toppings.	Often; it's good for you.	Several times a week.
21	**How often do you eat fermented foods (e.g., yogurt, sauerkraut, kimchi)?**	Rarely—except for sauerkraut on a hot dog.	Once in a while—not regularly.	A couple of times a week.	Regularly. I know it helps with digestion.
22	**Do you pay attention to the amount of fiber in your diet?**	No, not so much.	I eat vegetables, though not regularly.	Yes, I eat enough vegetables and grains.	Yes, I eat plenty of leafy green vegetables, whole grains, and/or legumes.
23	**How often do you eat *lean* protein (e.g., fish, poultry, beef, eggs, or soy products like tofu)?**	I eat meat/chicken every day and quite a few eggs, too.	I don't eat much fish/tofu but eat plenty of chicken, beef, and eggs.	Most of the time.	I don't eat much chicken or beef but eat plenty of fish/tofu.
24	**Do you tend to prefer fish and chicken over red meat and pork?**	No, if I'm honest, I like beef or pork (bacon, ribs, deli meat) better than chicken or fish.	I eat mostly poultry; red meat or pork once in a while.	I eat red meat and/or pork much less frequently than fish and chicken.	I prefer fish or chicken over red meat and pork.
25	**How often do you eat red meat, pork, or high-fat dairy?**	I eat one or more of these foods on a daily basis.	Several times a week.	A couple of times a month.	Less than once a week.

PAGE TALLIES

A=	B=	C=	D=

		A	B	C	D
26	**How often do you eat commercially raised chicken, beef, and/or pork? (If vegan/vegetarian: "If I were to eat meat, I'd choose . . .")**	All the time. Organic meat is too expensive, and I'm not sure it's really that different.	Quite frequently. I'm not really picky about it.	Occasionally, if I don't have access to organic meat.	Rarely or never. I prefer organic pasture-raised meat.
27	**How many eggs do you eat in a typical week? (Take into account eggs contained in baked goods, custards, quiche, etc. If vegan/vegetarian: "If I were to eat eggs, I'd choose . . .")**	I eat two or more eggs a day.	I eat a couple of eggs three to four times a week (about six to eight eggs per week).	I eat a couple of eggs one to two times a week (about two to four eggs per week).	I usually have one or two eggs a week, preferably cage-free.
28	**What kind of cheese do you eat most often?**	Canned or sliced and chunked deli cheeses.	Packaged cheese, especially cheddar or American.	Fresh, creamy cheese, like a Brie or Camembert.	Dry, aged cheeses like Parmesan or goat cheese like Feta.
29	**Do you include unrefined, extra-virgin vegetable oils such as olive, flaxseed, or coconut in your diet?**	Rarely. I prefer butter or margarine.	Olive oil for sure. The other oils, not as much.	I use them quite often.	I use unrefined oils all the time.
30	**Are you concerned with trans fats (and do you check labels to avoid purchasing them)?**	Not really.	Not overly concerned, but I keep an eye on the labels.	I avoid them as much as I can.	I don't eat any foods that contain trans fats and stick to organic sources instead.

PAGE TALLIES

A=	B=	C=	D=

		A	B	C	D
31	**Do you cook with herbs (e.g., rosemary, sage, garlic) in place of salt?**	No, I use salt when cooking/ baking.	It depends on the recipe; I typically use both.	I frequently use herbs instead of salt.	Yes, I always cook with herbs in place of salt.
32	**How often do you eat garlic and onions, either raw or cooked?**	Rarely or never.	Sometimes, like a few times a month.	Frequently, like once a week or so.	Once a week or more.
33	**Do you ever use the spice turmeric or curry in your meals?**	What is it?	Sometimes. I could take it or leave it.	I like curry! I'll have some form of curry every other week or so.	I love turmeric and include it in many of my soups, stews, and curries.
34	**Do you typically add salt to your food?**	I add salt to most everything; food tastes bland without it.	A little bit on a meal every day.	I don't typically add salt after cooking.	I use very little salt, if any, and mostly while cooking.
35	**How often do you drink 8 or more glasses of plain water a day?**	I don't like water much. I'd rather have juice or soda or something more flavorful.	Sometimes. I know I should drink more water.	Most of the time.	All the time.
36	**What are you most likely to do when you feel thirsty?**	Chug an ice-cold soda or commercial beverage.	An iced tea or juice would be nice.	Drink cold water or other beverage.	Water's my go-to (preferably at room temperature)!
37	**How often do you drink juice and/or soda?**	Every day. I might have OJ in the morning and soda sometime during the day.	Often. I don't like water very much.	Sometimes, preferably Diet Coke.	Rarely or never.
38	**How likely are you to start the day with a cup of herbal tea?**	Very unlikely.	Not my preference, but I'm fine with it once in a while.	I drink tea quite frequently.	Very likely.

PAGE TALLIES

A=	B=	C=	D=

		A	B	C	D
39	**How often do you drink red wine?**	Hardly ever. I prefer beer.	Occasionally, at a restaurant.	A couple of times a week.	I typically drink one to two glasses of red wine a day.
40	**How often do you enjoy a cocktail or a glass of hard liquor?**	Can't wait for Friday's happy hour!	Most evenings.	Once in a while, mostly socially.	Only on special occasions.
41	**How often do you drink milk as a beverage?**	Just about every day.	Frequently, like two to three times a week.	Occasionally, as a special treat.	Rarely or never.
42	**On a daily basis, how many cups of coffee or caffeinated beverages do you have?**	Several.	Two or three.	I start my day with a tall cup of coffee or other caffeinated beverage, and might have another cup later in the day.	I have one cup of coffee or caffeinated beverage at the most.
43	**How do you take your coffee?**	Vanilla, caramel macchiato, commercial frappé (e.g., Frappuccino), lattes—any of these will do.	With cream and sugar or sweeteners like Splenda or Sweet'N Low.	With sugar or sweeteners like Splenda or Sweet'N Low.	Black. I might add a dash of milk or alternate dairy.
44	**Do you use honey, maple syrup, or stevia instead of sugar/other sweeteners?**	Nope, I'd rather use white sugar.	I use sweeteners like Splenda and Sweet'N Low most of the time.	Yes, as much as possible.	Yes, all the time.
45	**What are your favorite kinds of desserts?**	Rich desserts, like cheesecake or a chocolate brownie.	Cake or ice cream, depending on the day.	I like variety—but everything in moderation!	Small and intense, like a piece of dark chocolate.
46	**How often do you enjoy desserts like pie, cookies, or ice cream?**	Almost every day. I have a sweet tooth!	Frequently, like three to four times a week.	About once a week.	Rarely—everything in moderation.

PAGE TALLIES

A=	B=	C=	D=

		A	B	C	D
47	**How often do you eat commercial pies, doughnuts, pancakes, pastries, and/or muffins?**	Nearly every day.	Frequently, like two to three times a week.	Occasionally, like once or twice a month.	Rarely or never. When I eat dessert, it's fresh, whole food with organic ingredients.
48	**Did you grow up eating commercially packaged chocolates, pastries, and candies (e.g., Hershey's Kisses, Twizzlers, Pop-Tarts)?**	Definitely—and I still eat them!	I did. I still might have something like that once in a while.	I might have eaten them but never loved them.	I didn't grow up eating candy.
49	**Fresh vegetables and fruits, nuts, legumes, and whole grains, with smaller amounts of fish and eggs. Does this sound like your typical diet?**	Not at all.	I eat like this once in a while, but it isn't my typical diet.	Not my typical diet pattern, but I often eat that way.	Yes. This is my typical diet pattern.
50	**Does your diet include mostly meat, white breads or pastas, potatoes, cheese, and/or fried foods?**	They happen to be my favorites.	Frequently. I'm a meat-and-potato kind of person.	I don't eat much bread or pasta, but I do like meat and the occasional fries.	Those foods are not a main part of my diet.
51	**What are you most likely to have for breakfast if time is no object?**	Something along these lines: eggs, bacon or sausage, pancakes, buttered toast, hash browns.	Fresh muffins, scones, eggs.	I'm flexible—scrambled eggs are good, but so is oatmeal and fruit.	Something light, like oatmeal or whole-grain toast (unless I'm having a fruit-and-veggie smoothie).

PAGE TALLIES

A=	B=	C=	D=

		A	B	C	D
52	**Does your typical lunch consist of a deli sandwich (e.g., a packaged ham, Swiss, and mayo kind of sandwich) and maybe chips on the side?**	Yes, that's me.	Frequently, maybe two to three times a week.	Occasionally, if I don't have access to other foods.	Rarely or never. I'm more of a soup-and-salad kind of person.
53	**At a barbecue, what are you most likely to throw on the grill?**	Burgers, ribs, hot dogs, cheeseburgers.	Mostly meat, but some corn would be nice, too.	A nice piece of chicken.	Vegetables like zucchini, eggplant, tomatoes. Shrimp and fish, too.
54	**How often do you eat each day?**	Three meals or more a day and two to five snacks.	Three meals a day and two substantial snacks.	Three meals a day and light snacks.	Three meals a day and a light snack occasionally.
55	**Do you typically eat small portions of food (e.g., meat or fish portioned to the size of the palm of your hand; cheese portioned to the length of your pinky finger; fruit the size of a small apple)?**	I generally eat larger portions of food.	Sometimes, when I'm not hungry.	Frequently—though I can eat larger portions of certain foods.	Most of the time if not always.
56	**How often do you eat larger portions of animal foods (e.g., more than 3 ounces of meat/poultry, more than 2 ounces of dairy, and/or more than two eggs)?**	Every day or almost. I prefer larger portions.	Quite often.	Occasionally. Mostly poultry and eggs.	Rarely or never.

PAGE TALLIES

A=	B=	C=	D=

		A	B	C	D
57	**How often do you realize you've eaten too much after a meal?**	Quite often! Sometimes I have to take antacids.	Often on weekends, and definitely on holidays.	I might end up eating too much over the holidays but usually have good control over what I eat.	Rarely or never.
58	**If the doctor told you to stop eating when you're 50 percent full, would you be okay doing that?**	How would I know when I'm 50 percent full?	It would be difficult, but I'd try.	Yes, but probably not every meal.	Not a problem. I never stuff myself.
59	**Are there at least twelve hours without any food between the time you finish dinner and the time you begin breakfast (e.g., dinner at 8 p.m., breakfast at 8 a.m., with no snacks in between)?**	Rarely. I usually have dessert or a snack later at night or before bed.	Sometimes. But more like six to eight hours instead of twelve.	Most of the time, but I'm not strict about it.	Yes, easily. I usually have an early dinner and refrain from eating until the next morning.
60	**Do you ever skip meals?**	Sometimes I don't eat from breakfast to happy hour.	I often skip breakfast.	Sometimes, if I'm in a rush or too busy to eat.	Rarely or never.
61	**How often do you feel like you need a pick-me-up after a meal (most likely something sugary or a cup of coffee)?**	I always have a cup of coffee or a piece of candy after lunch.	My blood sugar often drops early in the afternoon and I need a little pick-me-up.	Sometimes I'll have a cup of coffee or a little candy.	Rarely or never. I usually feel energized after a meal.

PAGE TALLIES

A=	B=	C=	D=

		A	B	C	D
62	**Do you cook your own meals or eat cooked-from-scratch meals made by someone else?**	I rarely cook. Or I might cook dinner if I have time.	Occasionally. I am happy with takeout and/or prepared foods.	I cook frequently. But I enjoy eating out, too.	I cook or eat meals prepared from scratch the vast majority of the time.
63	**How often do you eat processed foods like canned soups, frozen or prepared meals, commercial pastries/cakes, crackers, etc.?**	Often, like three to four times a week or more.	Frequently, like once a week.	Occasionally, like once or twice a month.	Rarely or never.
64	**When you're tired and coming home late from work, how likely are you to just pop a prepared dinner in the microwave (or eat any other form of prepared or packaged foods)?**	Very likely.	It could happen, but not all the time.	Quite unlikely, unless I have nothing else in the house.	Very unlikely.
65	**How often do you eat fast food and fried food?**	I love French fries and fast food. I eat them more than three times a week.	Fairly often—maybe three times a week.	Occasionally, like once a week.	Never, or once in a blue moon.
66	**How often do you eat low-fat or fat-free foods?**	Almost every day.	Frequently, like two to three times a week.	Occasionally, if I don't have access to other foods.	Rarely or never.

PAGE TALLIES

A=	B=	C=	D=

		A	B	C	D
67	**How often do you eat commercial breakfast cereals (e.g., Special K, Rice Krispies, Corn Pops, Cocoa Puffs, Raisin Bran)?**	That's an everyday breakfast for me.	Frequently, like two to three times a week.	Occasionally, like once or twice a month.	Rarely or never.
68	**How often do you find yourself reaching for a snack while you're at work or doing other activities, such as driving, reading, running errands, etc.?**	All the time.	Quite often. I'm a nibbler.	Once in a while, especially when doing something boring.	Rarely or never.
69	**What is your most typical snack?**	Chips, pretzels, candy, or cheese.	Candy or something sweet.	Nuts or nut butter.	A piece of fruit, some almonds, and/or yogurt.
70	**Do you snack after dinner or during the evening/night hours?**	Yes, I enjoy eating popcorn or chips while watching TV.	I often feel like a little dessert after meals, especially after dinner.	Occasionally. I could take it or leave it.	Rarely or never.
71	**When you go to the movies, how likely are you to get soda and popcorn?**	That's what the movies are about.	Popcorn, pretty likely.	Popcorn, occasionally.	Not my thing.
72	**What kind of peanut butter (or other nut butters) do you like best?**	One of the top commercial brands (Skippy, Jif, Peter Pan, etc.).	I usually buy flavored peanut butter (e.g., with salt, honey, chocolate).	I buy organic peanut butter as often as I can.	Organic and unsalted. I even make it myself sometimes.

PAGE TALLIES

A=	B=	C=	D=

		A	B	C	D
73	**How would you describe your digestion? How frequently are you constipated or have diarrhea?**	My digestion is not great. I often experience constipation or have diarrhea.	I feel constipated or have diarrhea with some frequency.	I'm usually regular, though once in a while I'll have a difficult day.	I am very regular. It's very rare that I experience constipation or diarrhea.
74	**After eating a meal, how often do you feel bloated, gassy, or uncomfortable?**	I have all these kinds of digestive issues most of the time.	Often. I frequently have a sensitive stomach (sometimes queasy or acidic).	Occasionally, mostly as a reaction to specific foods.	Rarely or never.
75	**Do you need to lose weight around your waist and hips?**	Alas, I do.	Yes, some. I tend to put on weight around my waist and hips.	Maybe a little bit.	Not really, I'm pretty fit.
76	**Do you gain or lose weight easily?**	I've been trying to lose weight forever.	I can gain a few pounds easily and have difficulty losing them.	I've been on a few yo-yo diets, but otherwise my weight is quite stable.	I've been my current weight for most of my life, give or take a few pounds.
77	**If you have a sore throat or feel like you're coming down with a cold or fever, how likely are you to drink plenty of water and get extra rest, rather than taking antibiotics?**	I'd rather get an antibiotic from my doc and get it over with.	Quite likely, depending on how bad I feel.	I might get a prescription anyway in case it gets worse.	I'm able to heal most things via good self-care. Antibiotics are a last resort.
78	**How frequently have you taken antibiotics since you were a child?**	More than once a year.	About once a year.	Less than once a year.	Rarely or never.

PAGE TALLIES

A=	B=	C=	D=

		A	B	C	D
79	**Is your blood pressure high? (If on medication, are you taking blood-pressure-lowering medications?)**	Yes, I have high blood pressure and/or I am taking blood-pressure-lowering medications.	My blood pressure is borderline high and/or I am considering taking blood-pressure-lowering medications.	It could be, if I didn't exercise and eat well.	My blood pressure is normal and I don't take blood-pressure-lowering medications.
80	**Do you have high cholesterol? (If on medication, are you taking cholesterol-lowering medication?)**	Yes, I have high cholesterol and/or I am taking cholesterol-lowering medications.	My cholesterol is borderline high and/or I am considering taking cholesterol-lowering medications.	I need to check my cholesterol, but overall it is within normal limits.	My cholesterol is normal and I don't take cholesterol-lowering medications.

PAGE TALLIES

A=	B=	C=	D=

SCORE YOUR TEST

Congratulations on completing the test!

All you need to do now is tally your scores. It's very simple. Just follow the four easy steps below.

1. On each page of the test, add up the number of times you circled choices A, B, C, and D, and write each subtotal at the bottom of the page in the Page Tallies box.
2. Add up the subtotals on each page and write the total for each column (A, B, C, and D) in the scoring box on the next page in the "Number of answers" row.

Part I—Diet	Total A answers		Total B answers		Total C answers		Total D answers		TOTAL
Number of answers									
Points per answer	0		1		2		3		
Total		+		+		+		=	

3. Score your answers using this scoring system:

- Assign 0 points for each A answer (so the total for all answers in column A will always be zero).
- Assign 1 point for each B answer (if you answered B ten times, the total is 10 x 1 = 10).
- Assign 2 points for each C answer (if you answered C ten times, the total is 10 x 2 = 20).
- Assign 3 points for each D answer (if you answered D ten times, the total is 10 x 3 = 30).

Add the totals for each column together. You will obtain a total score.

For example, my friend Lauren took the test and here are her results. She had 5 A answers (= 0 points), 10 B answers (10 x 1 = 10 points), 25 C answers (25 x 2 = 50 points), and 40 D answers (40 x 3 = 120 points). As you can see in the table on the next page, her total score is 180.

Part I—Diet	Total A answers		Total B answers		Total C answers		Total D answers		TOTAL
Number of answers	5		10		25		40		80
Points per answer	0		1		2		3		
Total =	0	+	10	+	50	+	120	=	180

4. Next, refer to your total score and select your classification using the following criteria:

- If your total score is lower than 80, you are at the Beginner level.
- If your total score is between 80 and 160, you are at the Intermediate level.
- If your total score is higher than 160, you are at the Advanced level.

This is your primary type within your unique constitution, which reflects the dominant characteristic of your overall brain makeup. For example, Lauren turned out to be at the Advanced level.

However, depending on how close you score to the cutoffs (scores around 80 and/or around 160), your profile might fall somewhere in between two levels. For instance, Lauren scored 180, which brings her closer to the next level down, the Intermediate level. As such, Lauren is at the Advanced level with some traits of an Intermediate level. Other people might fall somewhere in between Intermediate and Beginner levels. If that's you, too, make sure you read the recommendations for both levels.

NEXT STEPS

Once you have identified which level you are at—Beginner, Intermediate, or Advanced—move on to the next chapter to find out the meaning behind these levels, along with customized recommendations that make specific sense to your level in particular. This will give you a baseline to put the information in this book to use right away. You can start by using the very next chapter for some immediate tips on how to change or perfect your diet.

Try following your plan for at least three to four weeks and see how you feel. As you follow your recommendations, you will further refine the broader guidelines outlined in Step 2: Eating for Cognitive Power to unlock the maximum benefit for your individual needs. In fact, via this final step of discovering and embracing your own customized care, you will have integrated all the necessary information to craft an optimal brain fitness plan that is good for *you*.

At first, it might be difficult to eat and drink and be as active as the plan recommends. The first two weeks are usually the most challenging, so if you can make it through them, bravo! If you can't, don't worry. Old habits die hard, and everybody slips once in a while. The important thing is to come back to your plan and keep on going. Little by little, you'll be able to master your plan completely.

Once you feel that following the plan is going smoothly, try taking the test again. Most likely you'll find that your habits have improved and your level has shifted. If that happens, it's time to try the plan for your *new* level. Every level represents a step forward toward the optimal brain diet. We'll use this staging system to ensure that, over time, eating right for your brain becomes an integral part of your lifestyle, while making the transition as easy as possible.

Keep in mind that this approach does not offer a "magic pill" or a quick fix (and just as quickly lost) solution. The knowledge you discover practicing neuro-nutrition is the beginning of a lifelong journey. The goal here is to help you develop the healthiest brain you can possibly have, and to provide you with a map to keep it healthy for years to come. This goal is far more ambitious than shedding a few pounds for a one-time occasion and being done with it. So from your end, you need to be willing to challenge yourself, to go beyond your day-to-day routines, and to stay true to your own unique path to health. The ultimate goal is to help you get from any of the other levels into Advanced, and then to score at the top of the test, until you become your own personal brand of the ultimate centenarian. As with the actual centenarians living among us, your mission is to improve your lifestyle to maximize the longest, healthiest, most memory-rich life possible.

So . . . which level are you?

15

The Three Levels of Neuro-Nutrition Care

BEGINNER LEVEL

By and large, a brain-healthy diet is simply not the Beginner's cup of tea.

Among all the levels we'll cover, Beginners typically consume the largest amount of unhealthy foods laced with hidden sources of pro-inflammatory nutrients, trans fats, refined sugars, and harmful chemicals. At the same time, they rarely eat brain-essential, nutritionally dense foods like fresh vegetables, fruits, legumes, and/or whole grains—let alone fish or shellfish.

The recommendations for your level are meant to challenge you to leave your comfort zone and modify your eating habits to go easier on your brain. The plan outlined in this chapter addresses the above concerns and offers some practical solutions to achieve your goals for a long and healthy life. While nobody becomes nutritionally educated or physically fit with a snap of their fingers, what you can do is adopt as many of the recommendations as possible and channel them, little by little, into your daily routine. For instance, eating some vegetables is better than eating no vegetables. Drinking a glass of plain water is better than drinking a bottle of soda. Eating a cup of berries is better than eating a banana, but if you crave a banana, that's still better than a doughnut.

Our first goal is to reverse any shortage of vitamins, minerals, and fiber by increasing both your intake and diversity of fresh vegetables, fruits, legumes, and whole grains. In addition to the general guidelines outlined in Step 2: Eating for Cognitive Power, your plan includes specific recommendations as to which food items you should favor as a regular feature of day-to-day life.

Fruits, Veggies, Nuts, and Seeds

First and foremost, make an effort to eat vegetables and fruit *every single* day. Keep in mind that one of the major health factors hard at work in the centenarians' diet is their consumption of unprocessed rather than processed foods. Since centenarian islanders eat produce from their gardens and fields, they consume fewer pesticides and more nutrients, especially natural antioxidants. In keeping with this principle, make an effort to replace bland and nutritionally empty iceberg lettuce with a mix of fresh country greens and crisp, organic broccoli or cauliflower. And did you know that there are more than three thousand varieties of heirloom tomatoes in active cultivation worldwide? Next time you're about to buy the usual pale-looking beef tomato, try another variety instead—red, yellow, green, even purple. The brighter the color, the better it is for you.

As you kick off this level, your focus will be on eating at least one serving of leafy green vegetables (collard greens, Swiss chard, kale, or spinach) and/or cruciferous vegetables (broccoli, cauliflower, Brussels sprouts, or cabbage) every day. This is key, so really make a point of it.

Other veggies to keep on your radar are sweet peas (rich in omega-3s) and orange vegetables like carrots, winter squash, and butternut squash, which are rich in antioxidants and natural sugars. Last, it is important to include onions, garlic, and fresh herbs like sage and rosemary as a daily addition to whatever you're eating.

Remember that frozen, canned, and otherwise processed produce doesn't contain nearly the same amounts of brain-essential nutrients as does fresh, organic produce. If you don't have access to fresh produce, frozen organic options are still better than the commercially grown GMO varieties on most supermarket shelves. Your assignment is to make sure that you eat at least 1 cup of veggies with both your lunch *and* dinner.

Now let's talk fruit. Fresh berries such as blueberries, raspberries, blackberries, and strawberries, as well as citrus fruit like oranges, lemons, and grapefruit, are the power fruits you want to eat throughout your week. These fruits have a low GI and a good amount of fiber, in addition to being loaded with vitamins and antioxidants. One serving a day is a great start. As you will notice, bananas are not on our list, but if you are craving one, we'll make a once-a-week exception thanks to the brain-healthy glucose they contain. If you are a banana fan, experiment replacing them with berries and citrus over time. Apples are another good low GI choice—you know how the saying goes. While all apples are quite rich in vitamins, antioxidants, and fiber, Red Delicious and Gala apples have the highest antioxidant content of all and are therefore better for your brain.

Nuts and seeds are also important to include in your diet. The key is to focus only on *raw* nuts and seeds and to stay away from salted, roasted, spiced, honey-roasted, or otherwise sugar-coated and over-processed varieties. Start by making almonds and walnuts (skin on, if possible) a regular part of your snacks. As for seeds, try sprinkling a teaspoon of flaxseeds or a pinch of sunflower seeds into your soups and salads or yogurts and cereals. Flaxseeds are an excellent source of omega-3s, while sunflower seeds are the richest natural source of immune-supportive zinc.

If you're a peanut butter junkie, that's fine, but it's high time that you forgo commercial varieties like Skippy or Jif in favor of organic peanut butter. It isn't much more expensive and it is much better for you. The next step will be shifting from peanut to almond butter—a whole new world of flavor.

Grains and Legumes

When it comes to complex carbs, sweet potatoes top your list. These spuds are packed with antioxidant beta-carotene and vitamin C, and are a great source of vitamin B6, minerals, and fiber. They'll fill you up with goodness while cutting cravings for sweets at the same time. Make sure you eat a portion of sweet potatoes two to three times a week and as a replacement for the usual yellow or white ones.

Whole grains are another essential source of time-released energy. Make sure you eat one serving of whole grains, such as oats, whole wheat, and brown rice, twice a day. For example, this could be a slice of multigrain bread or a comforting bowl of oatmeal for breakfast and a cup of brown rice with lunch. Additionally, try to incorporate two or more servings of legumes like chickpeas or lentils every week. And don't forget to try my beloved Buddha bowls! Some recommendations to get you started are included in the sample menu plan at the end of this chapter.

Fish

For those of you out there who aren't crazy about fish, one way or another we need to provide your brain with enough omega-3s to improve your cognitive fitness and fend off diseases such as Alzheimer's. Perhaps you could add some lox atop a whole-wheat bagel? How do you feel about fish and chips?

Fish and chips are a great example of how to transform a fairly unhealthy meal into a nutrition-rich, tasty one. My version uses (organic, whole-grain) crunchy pretzels in lieu of processed bread crumbs, and features a meaty tilapia fillet instead of the typical bland whitefish. Moreover, fragrant coconut oil is used as the perfect replacement to the heart attack–inducing, partially hydrogenated oils used at any fast-food chain. The result is a knockout Pretzel-Encrusted Tilapia Fillet that will make your entire family swoon. The recipe is in chapter 16 and on my website (www.lisamosconi.com).

If canned tuna fish is your main source of fish, that's okay for a start. But when it comes to tuna, remember to choose wild or pole-caught tuna. Next, be sure to buy it in water (you can always add a little extra-virgin olive oil yourself). Did you know that salmon comes in cans, too? Canned wild Alaskan salmon is fairly easy to find in any supermarket. A large 15-ounce can of Bumble Bee Premium Wild Salmon costs as little as $2.75. And guess what? Anchovies and sardines are also stocked everywhere, not to mention mackerel. In no time, that same supermarket run is offering you five easy fish choices instead of one. Your mission, should you choose to accept it, is to start with two servings of fish per week and work your way up to three.

Meat, Sweets, and Processed Foods

As your consumption of wholesome grains, legumes, and fish increases, your intake of processed and fried foods, as well as sweets, meat, and dairy, will inevitably go down. In particular, watch out for trans fat–loaded commercial doughnuts, cookies, crackers, muffins, pies, cakes, Cool Whip–like creams, spreads, processed cheeses, and candy. Needless to say, these foods are *not* healthy and will subsequently wreak havoc in your system. Convenience foods like frozen and microwavable entrees, ready-made meals, and frozen dinners and pizza are also on our blacklist. It is crucial to limit how frequently you eat these foods until you're down to—well, never. You didn't really think Ethan Hunt leapt off building tops on fries and cupcakes, right?

It is also very important to keep an eye on the quality of your meat by minimizing consumption of commercially raised and processed meats while increasing that of pastured free-range chicken, grass-fed beef, and of course, wild-caught fish. This will further reduce the amount of trans fat, saturated fat, and cholesterol in your diet while at the same time protecting your stomach from pesticides, pollutants, and other harmful toxins that often come with your meat. Keep in mind that those meats are a hotbed for antibiotic-resistant bacteria mixed with all sorts of chemicals that have no business being in a healthy body. While increasing inflammation and oxidative stress in your every organ, including the brain, they simultaneously destroy your microbiome and intestinal health in the process. If you really love your bacon, focus on organic, antibiotic-free pork. It is hardest to find clean, safe versions of pork, which must be carefully selected.

Okay, meat eaters and dairy lovers, there are tricks and tips to enjoying these foods that you need to keep in mind. The number one tip is to limit both frequency and portion size. At the Beginner level, you will gently replace any old standbys with healthier, more nutritionally dense choices to improve and support the health of your brain. Focus on drastically reducing your consumption of red meat (beef and pork) and up that of free-range chicken, hen, duck, and turkey—*and* their eggs. As a rule of thumb, shoot for 3 ounces of cooked meat twice a week, and a couple of eggs once a week. If you feel that you need more, limit it to 4 ounces of poultry and 3 eggs per week—while making a conscious effort to reduce consumption over time.

In other words, your typical week should include three full non-meat, non-egg days.

If you crave a steak or are low on iron, grass-fed beef is the way to go—but no more than once a week. Be sure to follow these simple rules: (1) Stick to the recommended serving size (3 ounces, cooked), (2) choose a lean beef cut, and (3) grill it! No frying or broiling. For indoor grilling, I have a stovetop griddle with deeply grooved surface to gather fats away. Once you're at it, toss some veggies on that grill. Zucchini; eggplant; portobello mushrooms; and green, yellow, and red peppers; even blanched sweet potatoes—they all grill effortlessly outdoors and indoors. Also, if you're used to dressing your meat, feel free to top it with a small dollop of (organic, full-fat) butter and/or a squeeze of lemon, but avoid all other sauces, dressings, or gravies.

Let's talk about cheese now. Pecorino, Parmesan, aged Cheddar, cottage cheese, Feta, and goat cheese are all good choices, but think of them as a treat rather than a regular snack. Pretend you're on an airplane and your lunch tray arrives. Remember that small piece of cheese that usually sits in the coffee cup with the crackers? That's the perfect size, 1–2 ounces. One such serving per week will do. If cheese is your weakness, let's make it two servings to start.

The only dairy products we should consume on a regular basis are yogurt and kefir, though only when plain and unsweetened. Since they're an excellent source of brain-essential nutrients and probiotics, you can have a cup every day. Eating yogurt on a regular basis is crucial to maintaining optimal GI functions, which in turn supports brain health. Your goal is to avoid artificially sweetened and flavored varieties. Focus on plain yogurt, full-fat and always organic.

Finally, make sure that you reduce consumption of pro-inflammatory foods rich in omega-6s. An easy way to do this is by minimizing consumption of vegetable oils such as corn, soy, safflower, and sunflower. These oils are also often used in commercial preparations like ready-to-eat meals and frozen dinners, so read the labels. Better alternatives include unrefined (extra-virgin or at least virgin) cold-pressed oils from olives, flaxseeds, coconut, and avocado. Especially for beginners, it is crucial to stay away from refined oils. Keep in mind that extra-virgin oils (i.e., obtained by cold pressure) are much richer in antioxidants than any refined oils (obtained by solvent extraction), which are virtually devoid of these precious anti-aging compounds.

At the same time, make an effort to replace oversugared and salty foods with lighter, healthier options. An organic fresh-baked bagel as a Sunday treat is totally fine, as is high-quality dark chocolate. If you are new to dark chocolate, and sweet, white, or milk chocolate is what you're used to, dark chocolate will be a transition for your taste buds. Let a 65 percent bittersweet chocolate bar lead you in the right direction.

The next time you're about to reach for a doughnut, cookie, or candy bar, here's what you're going to do instead: eat a piece of dark chocolate *and* a handful of almonds. Then if you are still in the mood for dessert, take a look at chapter 16 or my website (www.lisamosconi.com) for several healthy dessert recipes that will help make the transition much easier, like my freshly made Banana Almond Pancakes, Raffaello Coconut Butter Balls, and homemade Chocolate Blueberry Ice Cream. It takes no longer than ten minutes to make any of the recipes.

Water and Beverages

While you are improving your food choices, let's make sure you keep well hydrated at the same time. You *need* to drink water. Plain water, to be specific. Remember that by committing to drinking eight glasses of water a day, you can increase your focus and reaction time by as much as 30 percent. The more water you drink, the more you'll come to realize its health effects, and the better you will feel, mentally and physically.

If plain water tastes a little . . . well, plain, try fruit-infused water instead. Lemon water is a great start. All you need to do is mix a large glass of water with the juice of half a lemon. Feel free to sweeten it with honey or maple syrup if that makes it more palatable. The best time to have this drink is first thing in the morning. It will kick-start your metabolism and stimulate digestion, while supplying a good amount of vitamin C, making you feel energized and "lighter" as you start your day. Several other fruit-infused waters are included on my website (www.lisamosconi.com).

Additionally, I created a smoothie recipe custom-tailored for you—the Soothing Cacao Smoothie. Chocolaty and creamy, it features the antioxidant power of raw cacao, chia seeds, and goji berries combined with memory-boosting ginseng and refreshing coconut water. It's basically a milkshake that happens to be really good for you. Its ingredients are also designed to curb sugar cravings, which will help you stay clear of fast-food-style shakes and ditch the soda can along the way. Many beginners enjoy this smoothie on a regular basis as the perfect replacement for milk beverages, frappés, and ice creams. After just a few weeks of drinking it on a regular basis (while following your synergistic plan as directed), you will see a dramatic improvement in both your energy level and clarity of thinking.

Watching your salt intake is also very important, as too much sodium not only promotes dehydration, but it can also increase your blood pressure, a known risk factor for heart disease. Make sure that your food is naturally flavored with herbs, spices, and healthy oils. As a result, they won't require additional table salt. If you really can't help yourself, limit added salt to no more than 1 to 2 light pinches a day.

Be aware of stimulants such as caffeine, alcohol, and sugar, which can also dehydrate and weaken your body and brain. As described in chapter 11, there are several coffee-kicking options, from cacao tea to yerba maté tea. Have fun experimenting with these new alternatives, while also limiting your coffee to the recommended number of cups per day: 1 cup of espresso or 2 cups of American coffee. Incidentally, the best way to drink coffee is black and freshly brewed. If you usually sweeten your coffee, that's fine, but it's time to say *sayonara* to white sugar.

For beginners in particular, I recommend *raw* honey as your main line of defense as you break away from white sugar. Raw honey is a true superfood and one of my favorite natural sweeteners. It's packed with enzymes, antioxidants, and brain-essential vitamin B6, not to mention a number of important minerals. Together, these nutrients help to neutralize aging free radicals while promoting the growth of healthy gut bacteria. Most honeys found in regular supermarkets are pasteurized, which means they are not raw. The pasteurization process deprives the honey of the majority of its nutrients, reducing it to something that is as bad for you as refined white sugar! Look for local, raw honey at farmers' markets and directly from local beekeepers. The darker the honey, the richer the flavor and the greater the health benefits. One tablespoon of raw honey has only 60 calories, most of which are derived from brain-nurturing glucose. Besides, raw honey has less impact on your blood sugar levels than a banana.

If you are used to artificial sweeteners like Splenda or Sweet'N Low, and are accustomed to using packets, try stevia instead. Stevia is a plant native to South America that has long been used in traditional medicine to support healthy blood sugar and prompt weight loss. Today, it is available in liquid drops, packets, dissolvable tablets, and baking blends. Check the label to be sure that it is not accompanied by additional, unwanted processing ingredients or added sugar. Stevia itself has zero calories and no side effects. Just keep in mind that it's twice as sweet as sugar, so a little goes a long way.

Exercise and Physical Activity

Along with improving your dietary choices, it is essential that you vary your routine to include even a small amount of physical activity each and every week—or, even better, every day. If you are new to exercising, I would recommend the speed-walking regimen described in chapter 10. Start by walking twenty minutes a day. Do so at a speed slightly faster than your normal pace, as if you were in a hurry. Little by little, increase your walking speed and duration. Once you are comfortable speed walking for twenty minutes, bring it up to twenty-five. Once you are comfortable with twenty-five, bring it up to thirty. Do this until you reach forty minutes of nonstop brisk walking, three times per week. Clinical trials have shown that this simple routine is an effective anti-aging strategy, not only counteracting brain shrinkage but even *reverting* it.

Sample Menu

All recipes are included in chapter 16. For more brain-healthy recipes, and to learn more about the nutritional value of these recipes, visit my website (www.lisamosconi.com).

UPON WAKING

Glass of warm water and lemon juice

BREAKFAST

Cup of green tea with raw honey

1 organic dried prune

Greek yogurt parfait with almonds, pomegranate seeds, and raw honey

or

Sicilian Scrambled Eggs

MORNING SNACK

Soothing Cacao Smoothie (half)

Probiotic supplement

LUNCH

Lentil Dal with Spinach

Half a baked sweet potato with a teaspoon of unrefined coconut oil *or*

Sweet Potato Chickpea Buddha Bowl

Cup of coffee, cacao tea, or dandelion tea

AFTERNOON TEA

Soothing Cacao Smoothie (the other half)

Brain-Healthy Trail Mix

DINNER

Essential Vegetable Soup

Dad's Lemon-Roasted Chicken

or

Pretzel-Encrusted Tilapia Fillet

Green salad with vinaigrette dressing

Glass of red wine

BEFORE BEDTIME

Herbal tea or fruit-infused water

INTERMEDIATE LEVEL

At the Intermediate level, your diet is not necessarily unhealthy but isn't optimized for brain health. By and large, your diet is characterized by a moderate consumption of fruits, vegetables, and grains and good amounts of protein and fat. Sometimes you even make an effort to choose organic options such as wild-caught fish, free-range chicken (and their eggs and dairy), and grass-fed beef over conventionally raised meats. However, these foods are not consumed regularly or often enough. Moreover, processed foods and fried and fast foods, as well as commercially raised red meat and dairy, are also on your menu, although not consistently.

This chapter includes several tips on how to move your diet toward one that emphasizes long-term nutrition for your brain.

Even though developing the new brain-healthy habits we've discussed might not have been your first instinct, taking the time each day to link these new behaviors with the important reasons for them is your personal key to becoming more fully equipped with both the knowledge *and* the skills necessary to protect and nurture your brain's health.

While some of the recommendations might seem easy, others are probably not. First and foremost, quality over quantity. For example, eating vegetables is better than no vegetables—but eating *organic* vegetables is better than eating just any vegetables. Likewise, protein is important for a number of brain functions, but some protein sources are better for you than others. In this regard, your plan focuses on replacing red meat and dairy with more vegetable sources packed with protein, like quinoa and beans, as well as traditional sources such as fish and chicken. Although you might be used to these choices, I will ask that you treat your brain a little better when selecting them, promising to buy poultry and eggs that are pastured and fish that is wild-caught rather than farmed as much as possible.

Second, I am asking you to "push the envelope" and be a little more adventurous. Your diet plan focuses on increasing consumption of special anti-aging foods that might not be typical of your current menu. This includes fish like wild-caught salmon, sardines, and mackerel, as well as my favorite brain food—caviar. Additionally, Brazil nuts, the richest natural source of selenium known to man, are among these superfoods. Vitamin C powerhouses like goji berries and great sources of omega-3s like chia and flaxseeds also top your choices.

In addition to the general guidelines outlined in Step 2: Eating for Cognitive Power, your plan includes specific recommendations as to which food items you should favor as a regular feature of day-to-day life. The goal is for you to retake the test a few weeks or months from now and find that your good habits have increased from "Often" or "Quite often" to "All the time" or "Most of the time."

Finally, I encourage you to practice overnight fasting a few times a week (i.e., refrain from eating for ten to twelve consecutive hours between dinner and the following day's breakfast). According to recent research, this simple practice will grant both your brain and your body renewed strength and resilience.

Let's look at these recommendations in more detail.

Fruits, Veggies, Nuts, and Seeds

Eating fresh produce is crucial for all levels, but at the Intermediate level in particular, *organic* is the name of the game.

Your first goal is to introduce more organic greens like my *nonna*'s dandelion greens into your diet. These greens come with an arsenal of vitamins and minerals that brain cells need to do their job and do it well. Plus, in some studies, wild greens were found to have ten times as many antioxidants as red wine. Remember this: people who eat one to two servings of leafy greens a day have cognitive abilities comparable to those eleven years their junior.

Your second goal is to eat fresh, vibrant vegetables *every single day* for lunch and dinner. Since you're already at the Intermediate level, most likely you have the basics covered. So in addition to spinach, kale, and broccoli, here are a few more options to get you inspired: carrots, Brussels sprouts, baby kale, baby spinach, cabbage, cauliflower, onions, beets, beetroot, fennel, chicory, zucchini, eggplant, tomatoes, avocado, olives, garlic, ginger, and sweet potatoes.

Most likely, you're also used to eating low GI foods like berries, citrus fruits, and apples. In addition to these, recommended fruit choices include lesser-known berries such as boysenberries, gooseberries, and goji berries, which come with a plethora of antioxidants and fiber. One serving a day is a good rule of thumb. Berries can be consumed twice a day instead, as these fruits are packed with fiber and brain-essential nutrients without adding too much extra sugar in the process. Sweet fruits with an intermediate GI, like peaches and nectarines, are also good choices, as are glucose-rich figs, once in a while.

Most likely you already snack on nuts and seeds every so often. Your goal is to eat them more consistently throughout the week, focusing on raw nuts and seeds and avoiding flavored varieties. Almonds, black walnuts, English walnuts, Brazil nuts, and pistachios, as well as pumpkin, sesame, sunflower, chia, and flax seeds are all good for your brain. If you like your nuts roasted or toasted, toast them yourself. Buying them roasted promises all sorts of unhealthy oils and condiments used by commercial manufacturers. It takes but a pan and just a few minutes to avoid this and perfume your kitchen at the same time. Nuts and seeds are also a great way to give your morning oatmeal some oomph, and to give soups and salads extra crunch instead of using commercial croutons.

If peanut butter is your jam, remember to choose an organic brand with no added salt or sugar (you can always add a sprinkle of sea salt, and peanuts are naturally sweet to start with). Another healthy trick is to swap your peanut butter for almond butter every other sandwich. Sunflower seed and pistachio butters are other excellent, flavorful alternatives packed with brain-essential nutrients. Just remember, no more than 1 tablespoon per serving. As small as they might be, these foods are nutritional powerhouses and pack a hefty punch of calories, especially when concentrated as a butter.

Grains and Legumes

It is important to limit consumption of the refined grains often used in pasta, and gradually replace them with healthier, more nutritionally dense choices that will improve and support the health of your brain.

When it comes to complex carbs, ancient grains are the ones that will top your new grocery list. Try amaranth, buckwheat, millet, spelt, kamut, and quinoa, as well as products such as flour, bread, and pasta made with these grains. This is not to say that you shouldn't eat oats, brown rice, or whole wheat. We're focusing on these lesser-known options to add variety and flavor to your table. When served together with beans, chickpeas, and lentils, these foods are an excellent source of complete protein that rival even the best cuts of meat. A little over 1 cup of lentils equals 3 ounces of chicken, for example. Additionally, these plant foods are packed with healthy carbs, fiber, vitamins, and minerals that will fuel your brain while leaving your glycemic levels undisturbed. If you're concerned about gluten, several ancient grains such as amaranth, buckwheat, and millet are naturally gluten-free. So is wild rice.

Make sure you eat one serving of whole grains twice a day and two or more servings of legumes a week. To get you started, your plan features recipes like Cannellini Beans and Ancient Grains Soup *and* my long-time favorite Chickpeas Tikka Masala. Warmed by soothing herbs and spices, these recipes are renowned for their nourishing and anti-inflammatory properties. Hope you'll love them as much as I do. For more inspiration, feel free to look up the recipes posted on my website (www.lisamosconi.com).

Fish, Meat, and Dairy

Since your diet includes moderate amounts of animal products, it is important to refocus your attention on the actual type and quality of these products. Keep in mind that farm-raised fish and commercial meats, including various cuts of butchered meat, sliced deli cold cuts, and any sort of packaged meat products, are all breeding grounds for dangerous antibiotic-resistant bacteria, not to mention pesticides, pollutants, and other harmful toxins.

Whether or not fish is a favorite food of yours, it's very important to eat a serving of it no less than three times a week. This ensures that your brain will receive enough omega-3s to maintain optimal cognitive fitness and ward off diseases such as Alzheimer's. That said, I'm not talking about tuna salad or the occasional shrimp cocktail. You need to make sure that you eat enough fresh fish like salmon and yellowtail tuna, as well as the tasty *ikura* (salmon roe). Several other options are included in chapter 16.

Other good sources of lean protein are organic, pastured meats; free-range chicken, duck, and turkey; Cornish hens; pheasant; quail; and these birds' eggs. Milk and dairy products from grass-fed cows are also safer. The key is to limit consumption of these foods to just a few times a week, while watching portion size, too. Refer to chapter 13 to get a sense of how much to eat of these foods and how often.

If you really crave a steak or feel a little low on iron, duck meat is a much more interesting option. Just recently, my husband discovered duck bacon at our local farmers' market. Uncured, smoked duck bacon has half the amount of fat of pork bacon but far more flavor, more lean protein, and a meatier texture than "alternative" bacons, such as turkey or mushroom. Grilled crispy to perfection in a cast-iron griddle, it has quickly become our main red meat dish (which we eat no more than once or twice a month, to give you a sense of how much and how often). Cured meats, cold cuts, ham, and bacon are not included in your plan at all. These animal products are too often full of chemicals, toxins, and bacteria, and should be completely eliminated.

As far as cheese goes, I'd recommend focusing on cheeses from sheep and goat milk, as they are richer in PUFAs and calcium than cheese made from cow's milk. These include all goat cheeses, Pecorino, and Feta, as well as ricotta and many farm-fresh cheeses. Again, everything in moderation: 1 to 2 ounces of cheese a week is plenty. In truth, after adolescence, humans no longer need to eat dairy or drink milk. However, if you are used to an occasional glass of milk, try goat's milk. Personally, goat's milk is the only type of milk I drink. I add a splash to my cup of rooibos and rose herbal teas, but it's equally excellent with coffee or black tea.

That said, plain and unsweetened yogurt is the exception to our dairy rules. Given that it is an excellent source of brain-essential nutrients and probiotics, you can have up to a cup every day. Eating yogurt on a regular basis is crucial to maintaining optimal GI functions, which in turn supports brain health. It might not come as a surprise that I recommend goat's milk yogurt. This yogurt is easier to digest than cow's milk, and yogurt contains fewer allergens, and causes less inflammation. From a nutritional perspective, it is higher in calcium and fatty acids, which makes it an ideal choice to strengthen bones, reduce inflammation, and improve absorption of other nutrients. Goat's milk and its yogurt are at their absolute mildest when fresh. Soon afterward they begin to taste stronger, or "goaty," depending on how they're handled. I buy my goat's milk products at our local farmers' market and they are mild and delicious.

If calcium is an issue, talk to your doctor about how best to address that, while keeping in mind that several nondairy foods are excellent sources of calcium, too. For example, 1 cup of cooked kale contains almost as much calcium as a glass of milk, and so do 3 ounces of sardines.

Sugar, Salt, and Processed Foods

Although Intermediate types don't typically eat a lot of processed foods, it is important that you work toward completely eliminating these foods from your diet. While most likely you wouldn't reach for a frozen dinner unless absolutely unavoidable, other less obvious junk foods can also pose a threat. Watch out for candies, commercial cookies, doughnuts, muffins, or pies—even crackers. Unless organic and freshly baked, these foods are just not worth eating. Other foods that we don't typically think of as "processed" include instant oatmeal, fruit juices (*all* fruit juices that aren't fresh-pressed are processed), and all low-fat dairy including yogurt, cream cheese, and spreads.

You also want to make sure that you reduce consumption of other pro-inflammatory foods, especially oils rich in omega-6. These include vegetable oils such as corn, soy, safflower, and sunflower. These oils are often used in commercial preparations like ready-to-eat meals, takeout food, and frozen dinners, which is another good reason to reduce consumption of prepared foods. The best condiments are unrefined extra-virgin olive oil and unrefined coconut, avocado, and flaxseed oils. In addition, use as much apple cider or balsamic vinegar, tamari, brewer's yeast, and miso paste as you like. This should also help you watch your salt intake. Too much sodium can increase your risk of high blood pressure, increasing your risk of heart disease in turn. All the recipes included in your plan and in chapter 16 are naturally flavored with herbs, spices, and aromatic oils. As a result, they don't require additional table salt. If you need an extra pinch, limit added salt to no more than the tip of a teaspoon per day.

We will also work to replace added-sugar foods with lighter, healthier options. In fact, regardless of which level you're at, this is a great time to say *auf wiedersehen* to white sugar once and for all. Healthier alternatives include honey, maple syrup (the darker the better), and stevia. Personally, I recommend coconut sugar. Most people have heard about the benefits of coconut water and coconut milk, but coconut sugar deserves a special spot in the limelight, thanks to its low glycemic load and rich nutrient profile. Not only is it packed with antioxidant polyphenols, but it also possesses good amounts of iron, zinc, calcium, and potassium. Make sure it becomes your natural sweetener of choice.

If you have a bit of a sweet tooth, visit my website (www.lisamosconi.com) for several healthy treats and desserts to help make the transition easier. For example, freshly made Banana Almond Pancakes, Raffaello

Coconut Butter Balls, and a decadent Creamy Chocolate Pudding make for a nice weekend treat. You're also welcome to swap these recipes out for a fresh fruit plate when you're in the mood, or with a piece of organic dark chocolate.

When choosing your chocolate, consider its cacao content. Those who have come over to the dark side know that a 75 percent cacao bar is healthier than a 65 percent bar. They've also transcended its once bitter flavor to discover they actually like the taste of cacao more than that of sugar. Besides giving chocolate its rich flavor, cacao powder delivers flavonol-related health benefits. So the more of it, the merrier. If you do well with a more intense and sharper flavor profile, pick a chocolate that boasts at least 75 percent cacao with a moderate-to-low cocoa butter and sugar content. If that's a bit too bitter, look for one with a smidge more butter and sugar, but the same percentage of cacao. A small 1-ounce piece of 75 percent dark chocolate a day is highly recommended to replenish your antioxidant storages and as a little pick-me-up.

Water and Beverages

Remember to drink at least eight full glasses of fluids every day, such as plain water, fruit-infused water, and herbal tea. Research shows that drinking plain water has beneficial effects that drinking soda will never have. For instance, drinking water before a test increases reaction times, making you think faster and perform better.

Many people at the Intermediate level report drinking an energy drink or soda to recharge. If that's you, the next time try this healthy trick instead: apple cider vinegar. It's simple: just mix a tablespoon of apple cider vinegar and a squeeze of lemon juice in a glass of water. Feel free to sweeten it with some maple syrup if it's too tart for your taste. The wide-ranging benefits of apple cider vinegar cover everything from curing hiccups to alleviating cold symptoms to helping with weight loss. As far as you are concerned, though, this beverage helps boost energy levels and is an antidote to fatigue. Next time you're feeling too tired to do your squats, take down a shot of this apple cider drink to get back on track. The fewer sports drinks and sodas you have, the better for you and your brain.

Additionally for you I recommend the Spicy Berry Smoothie. This luscious smoothie provides loads of vitamin C from sweet apples and lemon juice, combined with the antioxidant power of açaí, goji, and blueberries. It will make you forget about candy in a heartbeat. An unexpected light touch of cayenne pepper, ginger, and turmeric will further kick-start your metabolism, ensuring that you spring into action. By providing your hungry brain with a bounty of essential nutrients and natural sugars, this smoothie is the perfect companion for your journey to enhanced cognitive power.

Also, remember to limit your coffee to the recommended number of cups per day: 1 cup of espresso or 2 cups of American coffee. Keep clear of any add-ons, such as flavorings and toppings for hot drinks including whipped cream, all syrups, chocolate chips, caramel, fudge, or other sauces. Go for a dash of milk, or better, almond milk instead. If you like the comforting flavor of coffee but are interested in cutting back, coffee-kicking alternatives such as cacao tea and dandelion tea are excellent choices.

Last but not least, don't forget your small glass of red wine.

Portion Size and Intermittent Fasting

Now that we've discussed your diet overall, I've got a secret weapon for you. When you're ready, I would like for you to try to incorporate some of the lessons that we learned from the centenarians, focusing your attention on careful portion sizes, varying meal size (larger lunches, smaller dinners), and taking a try at *intermittent fasting.*

First, try sizing down your meals as the day progresses. Start the day with a good breakfast followed by a good-sized lunch, and close the day by eating your smallest meal at dinner. By eating a good breakfast, you'll give your metabolism a jump start and will be in better control of any cravings throughout the day. This way you won't be hungry by mid-morning and will be less likely to engage in mindless nibbling and snacking.

Once you have mastered this process, try practicing intermittent fasting a few times a week. Here's a quick-start plan for you to test the waters. Have an early dinner around six p.m. and refrain from eating anything more until seven or eight a.m. the following morning. That's an easy way to achieve a twelve-hour fast without too much trouble. While not eating, you can drink as much water and herbal tea as you like. If easier, feel free to start with ten hours without food and work your way up to twelve. Once you get used to having an early dinner, overnight fasting is an easy habit to adopt. I started practicing overnight fasting several years ago and at this point I can go for fourteen hours without eating, provided that I can get a good breakfast afterward, and find that I'm more alert and clear-minded when I wake up in the morning. However, everyone's reaction to fasting is different, so talk it over with your doctor, be smart, and go slowly.

Exercise and Physical Activity

While we're working to optimize your diet, let's take the same magnifying glass to your level of physical activity. The bottom line is that only a two-pronged approach that combines both diet *and* exercise will allow you to foster and achieve optimal brain health and cognitive fitness.

The goal is for you to regularly engage in moderate-intensity exercise. As mentioned in chapter 13, moderate intensity means that your heart will beat faster than normal. At this raised rate, you can carry on a conversation, but you probably can't sing.

Two of the most researched forms of brain-healthy exercise are walking and cycling. But swimming is just as effective, while being much gentler on the joints, and I often recommend it. Swimming is an easy exercise to begin at any age. It is particularly good for anybody who has injuries or any sort of condition that might prevent them from high-impact exercise like running or jogging. This water-bound exercise is effective at increasing muscle mass, strength, and flexibility, while improving pulmonary function, reducing high blood pressure, and taking the sting out of arthritis pain.

I've witnessed these bonus effects on my father, who used to suffer from high blood pressure and arthritis in his joints. Years ago my dad started swimming. At age seventy-five, he swims an average of twenty laps in just thirty-five minutes—three times a week! What's even better is that his blood pressure is back to normal and he's successfully avoided hip surgery for several years in a row.

Whether you prefer to hit the gym, swim, or dance, find the routine that works best for you and make sure it becomes an integral part of your life. Shoot for a minimum of three weekly sessions, forty-five minutes each in duration. Research shows that this routine is effective at improving brain fitness as well as cardiovascular strength, slowing down the aging process as a result.

Finally, remember that it isn't only about diet and exercise. Stimulating your brain intellectually and socially is crucial to a well-rounded lifestyle, ensuring optimal brain health and cognitive performance at the same time.

Sample Menu

All recipes are included in chapter 16. For more brain-healthy recipes, and to learn more about the nutritional value of these recipes, visit my website (www.lisamosconi.com).

UPON WAKING

Glass of warm water with lemon and apple cider vinegar

BREAKFAST

Cup of ginger tea with raw coconut sugar (if needed)

Blackberry Banana Muffin

or

Scandinavian Bircher Muesli

MORNING SNACK

Spicy Berry Smoothie (half)

Probiotic supplement

LUNCH

Go-to Green Salad

Cannellini Beans and Ancient Grains Soup

or

Rainbow Buddha Bowl with Maple Tahini Dressing

Cup of coffee, cacao tea, or dandelion tea

AFTERNOON TEA

Spicy Berry Smoothie (the other half)

Peanut Butter Power Bites

DINNER

Basic Wild Alaskan Salmon

or

Chickpeas Tikka Masala

Steamed green beans

Sweet potato mash

Glass of red wine

BEFORE BEDTIME

Herbal tea or fruit-infused water

ADVANCED LEVEL

Congratulations! You're one of the few people whose diet is already "brain nutritious." Because of your diet, you are way ahead of the game when it comes to maintaining and maximizing your cognitive fitness. Ready for more ways to get an additional gain? The recommendations included in your plan are the most challenging I have to offer.

The key for your type is to optimize your lifestyle for maximum brain payload. Given your nature, you will pay even closer attention to the details learned from the world's centenarians, since you are already taking their advice to heart. If you think back to chapter 9, you'll remember how centenarian communities generally have access to exceptionally clean air and water, and that their menus contain noticeably less fat, animal protein, salt, and sugar than the standard American diet. According to some experts, part of the reason they live so long is their regular consumption of foods high in essential omega-3s and antioxidant vitamins.

As we've learned, some foods are simply more brain-essential than others. Your goal is to increase consumption of those foods that provide the very richest sources of nutrition to your hungry, thriving brain. Another priority is to try out some fasting techniques. With your determination, curiosity, and newly gained neuro-nutrition expertise, an ultimate optimization of your brain health is just within your reach.

In addition to the general guidelines outlined in Step 2: Eating for Cognitive Power, your plan includes recommendations specific to this quest that suggest food items you should incorporate on a daily basis, if you're not doing so already. Although you've scored high on the test included in chapter 14, the goal is for you to take it again every few weeks to experiment with how to tweak and top your scores.

Many people at this level are reluctant to take supplements. In principle, as long as you feed your body and brain the right foods and nutrients, you shouldn't have any reason to take supplements. While this might be true in most cases, it is important to sometimes consider the need for occasional supplementation, especially as we age or during times of higher stress. Refer to chapter 12 for a complete list. It is important that you discuss this directly with your doctor.

Fruit, Veggies, Nuts, and Seeds

From now on, your brain-healthy mantra shall be *organic, fresh, diverse.*

Although eating fresh, organic produce is crucial for all levels, you are already on board. Knowing that frozen, canned, and otherwise processed produce doesn't contain nearly the same amounts of brain-essential nutrients as does fresh, organic produce, you've probably already made the effort to avoid it. For your level in particular, I recommend you dig a little deeper, introducing more organic wild greens such as my *nonna*'s favored dandelion greens, as well as Swiss chard and watercress for starters. When you're done with those, explore mustard greens, red and green chards, baby kale, mizuna, arugula, frisée, radicchio, and any lettuces from red romaine to red oak leaf, and Lollo Rosa to Tango. For you, the wilder, the better. These greens come with that intense volume of vitamins and minerals that our brain cells live for. In some studies, fresh-harvested wild greens were found to have *ten times* as many antioxidants as red wine and green tea put together.

Additionally, spinach has a very high antioxidant capacity, followed by peppers and asparagus, collard greens, rainbow chard, beetroot, fennel, radishes, chicory—these are only some options to get you started. Your mission is to continue to eat your carefully selected vegetables on a daily basis at both lunch and dinner.

You're probably quite friendly with low GI fruits like berries, so by now you're ready to tackle more than the average blueberry. Blackberries, red currants, raspberries, boysenberries, gooseberries, goji berries, mulberries, and amla (Indian gooseberry), as well as tart and black cherries, will lend depth and flavor to your hunt. Not only are these fruits packed with fiber, but they're unrivaled when it comes to their antioxidant vitamin and flavonol content, without adding sugar in the process. For you in particular, I recommend organic blackberries, and better yet, look for wild harvested ones. You can find them at your local farmers' market toward the end of the summer, as well as in many health food stores. Contrary to popular belief, blackberries contain even more antioxidants than blueberries, which makes this fruit an essential part of your anti-aging diet plan.

Citrus fruit like oranges, lemons, and grapefruit are other great choices. I encourage you to consume them year-round, especially during the winter, when fresh berries are scarce. For something sweeter, glucose-rich plums are also a good choice. Naturally, I recommend Italian plums in particular. Sometimes called the Empress plum, this small, meaty, egg-shaped fruit with blue or purple skin and yellow flesh is very rich in antioxidants and incredibly satisfying. These are the very plums that are made into prunes. Whether fresh, grilled, or pureed, make sure you eat your plums in season, toward the end of the summer.

For fruit servings, once a day is a good rule of thumb (except for berries, which can be consumed twice a day).

Additionally, remember to increase consumption of as many anti-aging plant foods as you can. Brazil nuts, the richest natural source of selenium on the planet, are a good start. Antioxidant-rich raw cacao also figures prominently. Mesquite, a largely unknown natural sweetener, possesses a mild caramel flavor and is good at regulating blood pressure. It might be your new favorite. Vitamin C–packed "ninja" goji berries and omega-3 heroes like almonds, chia, and flaxseeds should continue to grace your plate.

Most likely, raw nuts and seeds are already among your favorite snacks. What about freshly ground nut butters? All it takes is a high-speed blender and you can make your own signature brand. Personally, I am obsessed with chocolate roasted-hazelnut butter. Besides being delicious, hazelnuts are marble-sized superfoods that pack a potent nutritional punch. To make the butter, the trick is to remove the skin—which is easily done by bundling the roasted nuts in a towel and jiggling them around so that they grind against one another. Once that's done, the hardest part is behind you. Transfer the nuts to your food processor or blender and pulse until they become a velvety cream in about four to five minutes. I then add dark chocolate chips (65 percent cacao or higher) and pulse for a few more seconds. You can also add a pinch of sea salt to enhance the natural sweetness of this butter.

As long as you're not overdoing portion size, hazelnuts provide filling protein, fiber, unsaturated fat, and many essential vitamins and minerals like magnesium, B vitamins, and the antioxidant vitamin E. Shoot for 2 tablespoons of butter or a handful of nuts about three times a week.

Grains, Legumes, and Sweet Potatoes

I've made the Okinawan sweet potato a main staple of your menu. These bright purple powerhouses play a key role in the longer, healthier lives of Japanese centenarians. Also known as *beni imo*, the Okinawan sweet potato is even sweeter than its Western cousins, thanks to its higher glucose content. Moreover, a medium spud alone contains an incredible 500 percent of your daily dose of vitamin A, delivering a powerful punch of vitamin C, manganese, and fiber along with it.

Whole grains and legumes are other essential complex carbs to be consumed on a daily basis. Always focus on organic, unprocessed, and unrefined. In their raw state, these plant foods are packed with glucose and fiber, and are superb sources of vitamins and minerals. They will fuel your brain without triggering unpleasant sugar crashes in the process.

Make sure you eat at least one serving of whole grains not once but twice a day, and two or more servings of legumes a week. For some of you, this might be new. To help you along, your plan features recipes like the Winter Buddha Bowl, topped with a mouthwatering lemon sunflower sauce, and Avocado Toast . . . with a twist. We'll use Ezekiel bread instead of your regular bread. If you're not yet familiar with it, this bread is made of several different types of sprouted grains and legumes, typically wheat, millet, barley, spelt, soybeans, and lentils, making it truly distinctive, both in terms of nutritional content and its yum factor. Many other recipes are included in chapter 16 and on my website (www.lisamosconi.com).

Fish and Meat

As far as other proteins go, my recommendation is that you focus on them in this order: choose wild-caught fish as your main source of protein, followed by pastured eggs, organic grass-fed dairy, and lean and clean poultry.

Omega-3-rich fatty fish like salmon, sardines, anchovies, and mackerel should be a regular feature of your diet. In particular, I recommend wild Alaskan sockeye. Also known as "red salmon," this fish is treasured by many a chef and gourmand for its robust flavor, its high omega-3 content, and the bright rosy hue it keeps even after cooking. Additionally, being already at the Advanced level, you will appreciate what I consider the *ultimate* brain food—black caviar. Often thought of as a luxury gourmet food, black caviar consists of salt-cured sturgeon eggs. In addition to being incredibly rich in brain-healthy fats, black caviar contains extremely high levels of antioxidant vitamins and minerals, combined with a good amount of B vitamins and essential amino acids. It takes no more than 2 to 3 teaspoons of caviar to reach your brain-healthy DHA *and* choline doses for the day. Sprinkle some on rice crackers for a quick and pretty snack, or on whole-grain toast atop a dollop of Greek yogurt to serve it up as a hors d'oeuvre. Invest in a small jar that shouldn't cost more than $20 and see if it works for you. I buy mine at a Middle Eastern market, Sahadi's, in Brooklyn, but you can find it at any Russian deli or even online.

Eat as much fish and shellfish as you like, and limit (free-range) poultry such as chicken, turkey, quail, and their eggs.

As for dairy, plain, unsweetened yogurt is an exception to the not-too-much-dairy rule. Since yogurt is an excellent source of brain-essential nutrients and probiotics, you can feel free to have a cup every day. Try goat's and sheep's yogurt. They have an incredible nutritional profile and a more interesting flavor as compared to cow's milk yogurt, while also being easier to digest. Eating yogurt on a regular basis is crucial to a healthy digestion, which in turn supports your brain's health. A small piece of organic full-fat cheese once a week or so can be good for you, too, especially goat cheese or dry cheeses such as Pecorino.

Another important step in taking care of both your brain and gut health is to completely eliminate commercially raised meats from your diet. Cured meats, ham, and pork products like bacon, cold cuts, and packaged meats are chockful of chemicals, toxins, and even bacteria—and have no place in a healthy body like yours.

Also, review your consumption of healthy oils. You are probably already familiar with unrefined extra-virgin olive oil and cold-pressed flaxseed, coconut, and avocado oils. Macadamia oil, anyone? Hazelnut oil? It's time to explore all available options. At your level in particular, I recommend hemp oil. Hemp oil is a staple in the Bama Yao centenarians' diet. This oil is very rich in essential fatty acids as well as antioxidant vitamin E. Its ratio of omega-6 to omega-3 is about 3:1. This favorable ratio provides anti-inflammatory health benefits that might help to compensate for the general overconsumption of omega-6s typical to the Western diet.

Being mindful of these oils' smoke points, and when to eat them raw versus cooking with them, is also important (see chapter 12).

Sugar, Salt, and Processed Foods

Although you typically avoid processed and trans fat–loaded foods, it is important to make a conscious effort to completely eliminate these foods from your diet. While you probably don't use many prepared foods in your home, other less obvious junk food can sometimes sneak into your diet, posing a threat. Watch out for commercial candy, cookies, doughnuts, muffins, pies, and spreads—even crackers are not as innocent as they look. Unless organic and freshly made, these foods are not worth eating and you're just the type that can ultimately manage to get around the issue with careful planning.

Dark chocolate, instead, is always worth eating. As some of you might have noticed, chocolate makes you happy. It's not your imagination—it's a scientific fact! The key, though, is to focus on high-quality dark chocolate with high cocoa content and little or no added sugar. So when it comes to chocolate, the darker the better. Since you are at the Advanced level, you might already be used to eating dark chocolate with a cocoa content of 75 percent. I dare you to increase the cocoa content to 85 percent or 90 percent and see if you feel the difference. Many studies have confirmed that eating cocoa-rich dark chocolate benefits your brain, heart, and circulatory and nervous systems in one fell swoop. This is because the compounds present in this type of chocolate are effective at lowering insulin resistance, reducing blood pressure, increasing blood vessel elasticity, and combating inflammation.

Additionally, visit my website (www.lisamosconi.com) for several healthy, naturally sweetened treats, like Homemade Chocolate Ice Cream, Raffaello Coconut Butter Balls, and a variety of muffins and fruit-based desserts. Just to be clear, none of these recipes includes sugar.

In fact, if you haven't already, this is your moment to say a final *adiós* to white sugar once and for all, and slam the door shut on any use of artificial sweeteners such as Splenda, Sweet'N Low, or Equal. Healthier choices include raw honey, royal jelly, maple syrup, organic cane sugar, coconut sugar, brown rice syrup, and stevia, but also fruit butters (unsweetened apple butter in particular) and dried fruit like medjool dates. For your type in particular, I recommend trying the rarer natural sweeteners that also boast outstanding nutritional benefits, such as yacón syrup.

Yacón syrup is an up-and-coming superfood that is bound to cross your expert radar. This syrup comes from the root of a plant that grows in South America. Some people say it tastes like raisins, while others remark on the flavor being more like that of apples or caramel. Overall, it is quite sweet and goes well with many desserts. But the real benefit is that this natural sweetener is very high in oligosaccharides, a type of prebiotic sugar that feeds the good bacteria in your gut, thereby supporting your digestive health.

Finally, don't let salt get the best of you. Too much sodium can leave you feeling parched, while increasing your blood pressure at the same time. Focus on recipes that are naturally flavored with tasty herbs, spices, and healthy oils. As a result, additional table salt might be a moot point. If you need a touch of it, remember that white table salt is a no-no. Believe it or not, the table salt you think nothing of is a very processed food. It is refined to the point that most of its valuable minerals are stripped away and what's left is 99 percent sodium chloride. Harmful additives like aluminum hydroxide might also be added to prevent clumping. As we discussed, metals like aluminum are not safe, as they can deposit inside our brains and cause inflammation. Other purer, unrefined salts like sea rock salt or genuine pink Himalayan salt are much better for your brain, as they deliver more minerals without so much sodium, while being additive-free.

Coconut aminos, Bragg liquid aminos, tamari, miso paste, brewer's yeast, nutritional yeast, and all sorts of herbs and spices are other excellent alternatives.

Portion Size and 5:2 Fasting

Most people at the Advanced level don't typically overeat, nor do they snack late at night. They usually have an early dinner to leave enough time between digesting their food and going to bed and then refrain from eating until the next morning. In doing this, they have already incorporated some key centenarian habits into their routines.

Portion sizes and intermittent fasting are subtle but powerful tools in the centenarian's toolbox. Just in case you haven't experimented with this yet, let's go over the basics of these practices together. First, try sizing down your meals as the day progresses. Start the day with a good breakfast followed by a good-sized lunch, and end it by eating a smaller meal at dinner. As for portions, a quick guide to their sizes are as follows. When it comes to meat or fish, mimic the size of the palm of your hand (about 3 ounces). A serving of cheese? Think of a slice as the same length as your pinky. A serving of fruit should mirror the size of a small apple.

Once you have mastered this process, move on to intermittent fasting and practice it as often as you can. Here's how you start: Have an early dinner, around six p.m. or so, and refrain from eating until seven or eight a.m. the following morning. You can drink as much water and herbal teas as you like. You can achieve this twelve-hour fast without even trying, since you'll sleep through most of it. Research shows that overnight fasting is a good way to make your brain stronger and more resilient.

Once you're comfortable with overnight fasting, see if the 5:2 diet works for you. The 5:2 diet involves eating normally five days a week and eating no more than 600 calories a day for the remaining two days of the week. Check the sample menu starting on page 290 for a practical example. Recent studies show that this dietary regimen is effective at reducing inflammation and insulin resistance as well as lowering blood pressure levels, cholesterol, and triglycerides. It can be effective within just a few months' time. A note of caution: everyone's reaction to fasting is different, and fasting isn't for everyone. Be sure you discuss this practice with your physician before you start.

Exercise and Physical Activity

While you're perfecting your diet to maximize the health of your brain, it is also important to up your physical activity yet another notch. Aim for one to two hours a week of *vigorous* exercise, in tandem with your usual busy days.

If vigorous exercise is already a part of your weekly drill, there are still ways for you to add some variety and challenge to your activities to work your body in different and new ways. For example, there's a form of vigorous exercise that tends to fit your brain like a glove: hiking. A growing body of research shows that the human brain loves being outdoors. Outdoor exercise, even if moderate like a long walk in the park, has a soothing effect on the mind and has the power to literally improve one's mental health. This might even be a bit obvious from an evolutionary standpoint. We were born in the jungle, after all. But today most of us live in the urban jungle instead. Cities made of metal, glass, and cement rather than the green, natural spaces our species used to thrive in take their toll on us. As city dwellers, we spend far less time outside than our ancestors did, which has been linked to higher levels of stress, anxiety, depression, and other mental illnesses. Instead, our world's eldest and healthiest people don't live in polluted, cramped cities but cluster in smaller communities, along mountainsides and secluded rivers, immersing themselves in uncontaminated natural environments. I strongly believe that there must be something to that. Since breaking a sweat and being in Nature are both so good for your brain, it makes sense that doing them at the same time wouldn't be a bad idea. Make a point of walking uphill and without pausing for the first forty-five minutes to an hour to ensure aerobic exertion.

Besides, planning a long hike in the woods with your friends or family at least once a week is a great way to improve your loved ones' health at the same time. This will support the brain's need to be socially active, which is also crucial for long-term cognitive fitness.

Water and Beverages

Eight full glasses of fluids consisting of plain water, fruit-infused water, or herbal tea every day is the way to go. Also, I recommend that you try aloe vera juice to further hydrate your whole body.

Personally, every morning, at the start of my day, I drink a glass of water mixed with 1 fluid ounce of aloe vera juice, as well as a tablespoon of liquid chlorophyll. I highly recommend this potion. While you might know aloe vera for its well-established anti-inflammatory and hydrating properties, you might not be as familiar with chlorophyll. Typically marketed as an "internal deodorant" against bad breath and body odors, this ancient remedy has long been used to heal wounds, build new red blood cells, and improve blood oxygenation. This simple trick alone will stimulate your digestion while supporting your circulation, making you feel fresh and energized as you go about your day.

As you progress through the Advanced level plan, keep in mind another of my favorite supplements: noni juice. Every morning, my two-year-old daughter and I share a shot of noni juice, followed by the aloe drink mentioned above. I started doing this when she was eighteen months old, and at first, she would only take a sip. Then two sips. Now she requests her own shot every morning. This isn't the typical infant diet, of course. The knowledge I've gained in the lab has completely transformed our kitchen—and how I raise my daughter.

I discovered noni years ago when I was researching anti-aging foods. Because it hasn't been fully researched yet, you may not have heard of noni until now. I am happy to introduce you.

Noni (*Morinda citrifolia*) is an evergreen tree found from Southeastern Asia to Australia, especially in Polynesia. The juice is obtained from the noni fruit, which looks like an oversized mulberry or a small bumpy potato. Its flavor is about as appetizing as its appearance. Needless to say, most commercial brands mix it with juice, such as blueberry, to make it more palatable.

So what drew my attention to the lonely noni?

First of all, in many ways, I trust and value tradition. Noni juice has been used as medicine in the Pacific Islands for over two thousand years. Today it is still used for various skin problems, so much so that even supermodels are incorporating this wonderful fruit into their skin-care regime. Noni is also used to treat age-related ailments like arthritis and rheumatism, menstrual and abdominal pain, and even parasitic infections. When people keep drinking something despite its taste, there must be a good reason.

We now know that the beneficial effects of noni stem from its nutrient portfolio. Noni provides good amounts of vitamin C, potassium, magnesium, iron, zinc, and even some amino acids. But what makes this juice really powerful for brain health is that the antioxidants in noni are off the charts. Noni juice is surprisingly rich in antioxidants, from anthocyanins and flavonoids to beta-carotene, lutein, lycopene, and even selenium. These much-needed anti-aging nutrients work in synergy to prevent cell damage and reduce inflammation, especially as we age.

According to the National Institute of Health (https://nccih.nih.gov/health/noni), we don't know enough about noni to recommend that it be a regular part of our diets. In other words, it may or may not work for you. For me, it works well enough that I don't go a day without it. On top of the immediate energy kick and its positive effects on my skin, I find that noni juice stimulates digestion and supports healthy elimination, therefore boosting the immune system too. If you feel up for the challenge, online retailers carry several brands of Tahitian noni juice.

My suggestion is to buy the original undiluted and unsweetened version—and if you find it too bitter, add some honey or dark cherry or pomegranate concentrate yourself.

Additionally, I created a smoothie recipe custom-tailored for your level: the Nourishing Green Smoothie. This nutrient-dense smoothie is an antioxidant powerhouse thanks to its signature blend of chia, flaxseeds, and açaí and goji berries. These are combined with omega-3-rich almonds, nourishing goat's milk, and Nature's hydrator, unsweetened coconut water. And yes, it will practically glow green. This is because your smoothie will contain spirulina as well. Spirulina boasts the highest protein content of any food on the planet. It is also high in iron, many B vitamins, and vitamins A, C, D, and E. To give you a sense of its potency, just 1 tablespoon of this velvety, dark green powder provides 70 percent of your daily vitamin B12 and over 300 percent of vitamin A.

Finally, most people at the Advanced level don't drink caffeinated soda at all, nor do they overdo it with coffee. As a rule of thumb, I recommend no more than 1 cup of espresso or 2 cups of American coffee per day. Keep in mind that espresso has the highest anti-aging capacity among all beverages. If coffee gives you the jitters, tea is a great alternative. Just remember that green tea contains more antioxidants than black tea, and is therefore better for you—and your delicate brain.

Last but not least, raise your glass of organic red wine. If you don't drink alcohol, or just for a change, try pomegranate juice. As mentioned in chapter 11, pomegranate juice delivers a unique repertoire of tannins, anthocyanins, and unsaturated fatty acids. These substances are so powerful that some pomegranate juices, especially the concentrated ones, can have up to three times the antioxidant capacity of red wine and green tea combined. Cut it with a little water and a slice of lime for a fancy cocktail at parties.

Sample Menu

All recipes are included in chapter 16. For more brain-healthy recipes, visit my website (www.lisamosconi.com).

UPON WAKING

Glass of water with aloe vera juice and minty chlorophyll

BREAKFAST

Cup of ginger tea with lemon and raw honey

Cup of fresh berries (blueberries, strawberries, blackberries, et cetera)

Swiss overnight oats

or

Avocado Toast

MORNING SNACK

Nourishing Green Smoothie (half)

Probiotic supplement

LUNCH

Ayurvedic Mung Bean Soup

Steamed zucchini with extra-virgin olive oil

or

Winter Buddha Bowl

Cup of coffee, cacao tea, or dandelion tea

AFTERNOON TEA

Nourishing Green Smoothie (the other half)

Gala apple with almond butter

or

Black caviar over rice crackers

DINNER

Grilled Salmon in Ginger Garlic Marinade
Steamed broccoli
Cup of wild rice with coconut oil and tamari
or
Grilled Sweet Potatoes with Fresh Spinach Salad and Yogurt-Tahini Sauce
Nonna's Dandelion Greens with Lemon Juice and Extra-Virgin Olive Oil
Glass of red wine

BEFORE BEDTIME

Herbal tea or fruit-infused water

16

Brain-Boosting Recipes

BREAKFAST

Avocado Toast

If Avocado Toast isn't in your weekly routine, this recipe will change that. Nothing beats this classic toast. Start with a silky avocado, chockful of heart- and brain-healthy fats, and dress it with extra-virgin olive oil (or another heart/brain favorite like flaxseed oil) and a pinch of pink salt and chili pepper flakes to taste.

My personal neuro-nutritional touch is to replace commercial bread with Ezekiel bread. Ezekiel bread is prepared using traditional methods of soaking, sprouting, and baking that have been in existence for centuries. It contains several different types of sprouted grains and legumes, typically wheat, millet, barley, spelt, soybeans, and lentils. Since it is completely preservative-free, Ezekiel bread is best kept frozen. Since it is also completely free of any added sugars, this bread provides a good amount of brain-healthy glucose without a heavy glycemic load.

INGREDIENTS *(Serves 2)*

2 slices Ezekiel bread

1 ripe avocado

1 tablespoon fresh lemon juice

Pinch of Himalayan sea salt

Chili flakes

1 teaspoon extra-virgin olive oil

DIRECTIONS:

STEP 1. Toast the bread.

STEP 2. Cut the avocado in half, remove the pit, and scoop the flesh into a bowl. Add the lemon juice and sea salt to taste. Mash the ingredients together with a fork, keeping the texture slightly chunky.

STEP 3. Spread the mash onto each piece of toast and garnish with some chili flakes and a drizzle of extra-virgin olive oil. (If you prefer your avocado sliced instead of mashed, sprinkle on your toppings.)

Bircher Muesli

Bircher Muesli was developed toward the end of the nineteenth century by Swiss physician Maximilian Bircher-Brenner for his patients. It is still a very popular breakfast throughout Switzerland and Germany. Muesli (pronounced *muse-lee*) is an uncooked mixture of nuts, seeds, grains, dried fruits, and spices. Although the ingredients might remind you of granola, the main difference is that muesli doesn't contain any added oils or sugars, and is eaten raw.

This muesli is typically mixed with nut milk, yogurt, or fruit juice, in which it soaks overnight. Soaking reduces the amount of phytic acid present in whole grains, nuts, and seeds, making them more easily digestible. On top of that, soaking increases our bodies' capacity to absorb brain-essential minerals like zinc, iron, and also calcium.

There are infinite variations of the basic Bircher recipe. Below you'll find my top two choices. They all make for a light nutrient-dense breakfast, and are incredibly easy to make and serve. Making them ahead of time and storing them in the fridge will allow you to have a healthy breakfast at the ready every morning. They will last refrigerated in a sealed container for up to a week.

SCANDINAVIAN BIRCHER MUESLI

INGREDIENTS ***(Serves 8)***

FOR THE MUESLI:

1 cup organic steel-cut oats

⅔ cup wheat flakes

½ cup toasted wheat germ

1 tablespoon psyllium whole husks

¼ cup flaxseeds, ground

⅓ cup almonds and/or walnuts, coarsely chopped

2 tablespoons unsweetened coconut flakes

½ cup date nibs, or dried apricots or dried figs

2 tablespoons local, raw honey

1 teaspoon mesquite powder

Pinch of cinnamon

3 cups filtered water

2 cups organic plain yogurt (I prefer goat's milk yogurt)

DIRECTIONS:

STEP 1. In a medium-size bowl, mix together all the ingredients (except the yogurt) until well combined. Transfer the muesli to small glass jars. Let sit covered overnight.

STEP 2. In the morning, add the yogurt.

SWISS OVERNIGHT OATS (WITH A TWIST)

INGREDIENTS ***(Serves 4)***

FOR THE MUESLI:

2 cups organic steel-cut oats

1¾ cups organic grass-fed whole milk

¼ cup organic apple juice

3 tablespoons fresh lemon juice

1 apple, cored and grated with the peel

2 tablespoons local, raw honey

¼ cup raisins

1 tablespoon chia seeds

Dash of ground cinnamon

FOR THE TOPPINGS:

1½ cups organic plain yogurt

1 cup blueberries (other berries work well, too)

½ cup hazelnuts, chopped

DIRECTIONS:

STEP 1. In a medium-size bowl, mix together all the ingredients for the muesli until well combined. Transfer the muesli to small glass jars. Let sit covered overnight.

STEP 2. In the morning, add the yogurt and top with the blueberries and hazelnuts.

Blackberry Banana Muffins

Can muffins be good for you? Of course—if you know how to make them right.

This recipe is *packed* with brain-healthy nutrients. There are vitamin E and omega-3s from the almonds and walnuts. A bounty of essential minerals is supplied thanks to the chia seeds and honey. Oats lend the recipe a good amount of soluble fiber, while coconut oil supplies these muffins with healthy fats. Memory-boosting choline and serotonin-building tryptophan come from the organic eggs used, while vitamins A and C and a burst of antioxidants are delivered by the berries.

These muffins are quite rich and filling, so I typically use 12-cup muffin pans. As always: quality over quantity, and everything in moderation.

Enjoy your muffin with a cup of tea (chai goes well with this recipe) and some extra berries for a satisfying and cozy brain-healthy breakfast or snack.

INGREDIENTS ***(Serves 8)***

1½ cups organic rolled oats

5½ tablespoons extra-virgin coconut oil

1 cup organic almonds

½ cup organic walnuts, chopped

2 tablespoons chia seeds

2 teaspoons baking powder

¼ teaspoon baking soda

2 organic free-range eggs, plus 1 egg white

1 cup goat's milk yogurt

2 teaspoons ground cinnamon

Zest of ½ lemon

¼ cup local, raw honey

2 medium ripe bananas

1 cup organic blackberries

DIRECTIONS:

STEP 1. Preheat the oven to 350°F. Lightly grease a 12-cup muffin pan with 1½ tablespoons coconut oil.

STEP 2. Using a food processor or a high-speed blender, grind the oats and almonds to a flourlike texture. In a large bowl, combine the ground oats and almonds, the walnuts, chia seeds, baking powder, and baking soda.

STEP 3. In a separate bowl, lightly beat the eggs. Add the yogurt and stir until well combined.

STEP 4. Place the coconut oil, cinnamon, lemon zest, and honey in a small saucepan over very low heat and stir until the mixture reaches a syruplike consistency.

STEP 5. Add the egg mixture to the dry mixture and stir before adding the coconut-oil mixture.

STEP 6. Mash the bananas and add them to the bowl. Add the blackberries and stir gently.

STEP 7. Fill each cup three-quarters of the way full. Bake for 25 to 30 minutes or until the tops are golden brown. Allow to cool for 15 minutes before removing from the pan.

Sicilian Scrambled Eggs

A staple in many American diets, scrambled eggs make for a filling breakfast and also work well as part of a quick and easy lunch or dinner. Except for people sensitive to dietary cholesterol, most of us are no longer advised to avoid eating eggs, since consuming eggs in moderation isn't likely to increase blood cholesterol levels. "In moderation" means no more than one or two eggs per meal, and no more than once or twice a week (even if eggs are no problem for you).

Eggs offer protein and several brain-essential B vitamins like choline, B12, and B6, as well as anti-aging selenium, which make them a healthy addition to your diet. Mixing in other nutritious ingredients, such as spinach, tomatoes, and olives, further boosts their nutritional value. This quick and easy Italian recipe is loaded not only with nutrients but also with the robust flavor of fresh basil, pungent garlic, and extra-virgin olive oil.

INGREDIENTS *(Serves 4)*

6 organic free-range eggs

2 ripe plum tomatoes, diced

¼ cup Kalamata olives, diced

¼ cup organic grass-fed whole milk

½ cup Feta

1 tablespoon extra-virgin olive oil

2 cloves garlic, finely chopped

2 cups baby spinach

½ cup fresh basil

Sea salt and black pepper

DIRECTIONS:

STEP 1. In a medium bowl, whisk together the eggs, tomatoes, olives, milk, and Feta.

STEP 2. In a large pan, heat the olive oil over medium heat. Add the garlic and cook for 1 minute or until lightly brown. Add the spinach and basil and cook for 1 minute more.

STEP 3. Add the egg mixture to the pan and toss with a spatula for 2 to 3 minutes. The eggs should be tender but not runny. Add the salt and pepper to taste. Serve immediately.

ENTRÉES

Ayurvedic Mung Bean Soup

This traditional Ayurvedic recipe has been used all across the Asian continent as a healing medicine for thousands of years. Well renowned for its nourishing and detoxifying effects, mung bean soup makes for a delicious meal possessing anti-inflammatory properties that act as a tonic to the nervous system. A good source of fiber and soothing herbs like ginger, this recipe is excellent for those who have digestive issues such as bloating, gas, and constipation.

INGREDIENTS *(Serves 4)*

2 cups organic split mung beans

1 tablespoon extra-virgin coconut oil

1 teaspoon ground turmeric

Pinch of ground cumin

1 small yellow onion, thickly sliced

3 garlic cloves, minced

One 1-inch piece fresh ginger, grated

1 organic carrot, peeled and finely chopped

2 organic celery stalks, finely chopped

½ teaspoon dried rosemary

6 cups organic vegetable broth

Sea salt

DIRECTIONS:

STEP 1. Soak the beans in plenty of water for at least 5 hours or overnight (soaked beans are easier to digest and will not cause bloating).

STEP 2. Heat the oil in a large, heavy saucepan over medium heat. Add the turmeric and cumin and stir for 1 minute.

STEP 3. Add the onion and cook, stirring frequently, until golden and soft, about 5 minutes. Add the garlic and ginger and cook for 2 minutes more.

STEP 4. Add the carrots, celery, soaked mung beans, and rosemary. Stir to combine. Add the broth and bring to a boil. Reduce heat to low. Season with sea salt to taste. Cover and simmer for 25 minutes or until the beans are cooked but not mushy.

Basic Wild Alaskan Salmon

This recipe is another household favorite. My husband says this super-simple dish is among those that top his list.

The simplicity of the recipe will surprise you, as there is hardly any cooking involved. Both the delicious flavor and nutritional punch rely on using the freshest highest-quality ingredients available and the fact that they combine so well together.

What is brain-healthy about this recipe? Everything.

First, the fish. Deep-water fish, such as salmon, are rich in omega-3 polyunsaturated fatty acids, especially DHA, an essential for brain function. DHA makes up to 50 percent of the brain's phospholipids, which are crucial in keeping brain cell membranes flexible and functional as we age. Additionally, omega-3s have scientifically proven anti-inflammatory effects.

Seeing is believing, so I am treating my family to the best possible source of DHA: wild Alaskan salmon. Since it's wild-caught, wild Alaskan salmon is "cleaner" and healthier than farm-raised fish, loaded with omega-3 PUFAs and containing 22 grams of lean protein that includes all of your essential amino acids per just 7 grams of fat (most of which are essential fats)! All these bonuses in as little as 3 ounces of fish. The dressing here is important, too, as it really complements the dish.

INGREDIENTS ***(Serves 4)***

1 pound wild Alaskan salmon fillet, cut in 4 pieces (I use frozen wild Alaskan salmon when I don't have access to fresh)

¼ cup filtered water

2 tablespoons extra-virgin coconut oil

2 tablespoons tamari

Juice of ½ lemon

DIRECTIONS:

STEP 1. Rinse the fish and arrange the fillets, skin side down, in an enamel or glass pie plate. Add the water. Set the pie plate in a steamer, cover, and steam until the fish is cooked through, about 8 to 10 minutes.

You can also use a skillet.

Heat a skillet over low heat. Place the fillets skin side down on the skillet and add the water. Cover and let simmer for 4 to 5 minutes.

STEP 2. In a small saucepan, heat the coconut oil, tamari, and lemon juice and stir for 1 minute.

STEP 3. Transfer the fish to a plate and drizzle the tamari-lemon sauce over. Serve immediately. I usually serve this with a small cup of brown rice.

BUDDHA BOWLS

A Buddha bowl is so full of all things good that when served, it has the appearance of a rounded "belly" (much like the belly of a Buddha). Sometimes referred to as a glory or hippie bowl, Buddha bowls are hearty, filling dishes made of raw or roasted vegetables, legumes such as beans and lentils, and healthy grains like quinoa or brown rice. Depending upon the recipe you choose to follow, the dish can contain a rainbow of ingredients. Sometimes they also include toppings like nuts and seeds—and incredibly flavorful dressings. And the best part is, each of the Buddha bowl recipes listed below is simple to make and jam-packed with filling nutrients and vitamins that nourish and protect your brain.

Rainbow Buddha Bowl with Maple Tahini Dressing

This recipe is very filling and the dressing is the "frosting on the cake"—an adaptation of a savory sauce I ate at Life Alive, an "urban oasis and organic café" located in Cambridge, Massachusetts. My husband used to work at the Massachusetts Institute of Technology (MIT), and whenever I spent a weekend there, we would go to Life Alive and enjoy their signature dish, The Goddess, my inspiration for this Rainbow Buddha Bowl recipe.

As you can see, Buddha bowls take time to prepare, so be smart about it: make more and keep leftovers for another day!

INGREDIENTS *(Serves 4)*

FOR THE BOWL:

½ cup finely chopped baby kale (3 large leaves)

½ cup peeled and chopped carrots

½ cup peeled and chopped red beets

½ cup chopped broccoli

¼ cup slivered almonds

½ cup cubed firm tofu

1 cup cooked wild rice

½ cup cooked quinoa

FOR THE DRESSING:

2 small garlic cloves

One 2-inch piece fresh ginger, peeled

2 tablespoons tamari or nama shoyu sauce

Juice of ½ lemon

2 tablespoons organic tahini

1 tablespoon extra-virgin coconut oil

DIRECTIONS:

FOR THE BOWL:

STEP 1. Place the kale, carrots, beets, and broccoli in a steamer or large pot with one-quarter cup of water. Steam over medium heat until the vegetables reach desired texture, 2 to 4 minutes.

STEP 2. In a skillet over medium heat, toast the almonds for 1 minute, stirring constantly.

STEP 3. Combine the steamed vegetables, toasted almonds, and the rice and quinoa with tofu in a serving bowl.

FOR THE DRESSING:

Place all the ingredients in a food processor or high-speed blender and blend until creamy.

TO PREPARE THE BOWL:

Add the dressing to the bowl, mix together, serve, and enjoy.

Sweet Potato Chickpea Buddha Bowl

This recipe is loaded with four kinds of vegetables along with fiber- and protein-rich chickpeas, antioxidant-rich sweet potatoes, and healthy grains that pack a healthy B vitamin punch. All of this is finished off with a maple-tahini sauce that's so good you'll want to put it on *everything.*

INGREDIENTS ***(Serves 4)***

FOR THE BOWL:

2 medium sweet potatoes, halved (try using Okinawan sweet potatoes instead of regular yams)

½ red onion, sliced into wedges

2 tablespoons extra-virgin coconut or grapeseed oil

2 cups chopped broccolini (large stems removed)

1 cup sliced shiitake mushroom caps

¼ teaspoon Himalayan pink or kosher salt

1 cup cooked brown rice

2 cups baby spinach

2 cups mesclun greens

(continued)

FOR THE CHICKPEAS:

1 (15-ounce) can chickpeas, drained, rinsed, and patted dry

1 teaspoon ground cumin

½ teaspoon ground turmeric

1 teaspoon mustard seeds

2 garlic cloves, minced

Pinch of chili powder

Pinch of Himalayan pink salt

1 tablespoon coconut oil

FOR THE DRESSING:

¼ cup organic tahini

2 tablespoons organic maple syrup

Juice of ½ lemon

1 garlic clove

One 1-inch piece fresh ginger

2 to 4 tablespoons hot water

DIRECTIONS:

FOR THE VEGETABLES:

STEP 1. Preheat the oven to 400° F. Arrange the sweet potatoes, skin side down, and onions on a baking sheet. Drizzle with 1 tablespoon of the oil, making sure the flesh of the sweet potatoes is well coated.

STEP 2. Bake for 10 minutes, then remove from the oven, flip the sweet potatoes, and add the broccolini and mushrooms. Drizzle the vegetables with a bit more of the oil and season with the salt. Bake for 8 to 10 minutes more, until sweet potato is soft, then remove the pan from the oven and set aside.

FOR THE CHICKPEAS:

STEP 1. In a medium bowl, toss together the chickpeas and spices.

STEP 2. Heat the remaining coconut oil in a large skillet over medium heat. Once the oil is hot, add the chickpeas and sauté, stirring frequently, for about 10 minutes or until they are browned and fragrant. Remove from the heat and set aside.

FOR THE DRESSING:

Place all the ingredients in a high-speed blender and pulse for 1 minute to combine, adding hot water as necessary to thin the dressing. Set aside.

TO PREPARE THE BOWL:

Slice the sweet potatoes into bite-size pieces. Divide rice, vegetables, and greens among 4 bowls and top with chickpeas and tahini dressing. If you like, sprinkle chopped nuts and seeds on top.

Winter Buddha Bowl

This recipe is loaded with brain-healthy cruciferous vegetables, such as broccoli and cauliflower, along with fiber- and protein-rich kale, quinoa, and beans. All of this is finished off with a lemon-sunflower dressing that's almost too good to be true. Remember to add some pickled vegetables (in brine) to nourish and support your microbiome as well!

INGREDIENTS ***(Serves 4)***

FOR THE BOWL:

2 cups chopped broccoli florets

3 cups halved Brussels sprouts

1 cup chopped Tuscan kale

1 carrot, chopped

2 tablespoons extra-virgin olive oil

Sea salt

(continued)

FOR THE DRESSING:

1 cup cashews

2 small garlic cloves

½ cup fresh sage leaves

4 tablespoons hulled hemp seeds, plus more for sprinkling

2 tablespoons sunflower seed butter

1 teaspoon red miso paste

Juice of ½ lemon

3 tablespoons tamari

1 teaspoon local, raw honey

¼ cup water

1 cup cooked spelt grains

1 cup cooked buckwheat

1 cup cooked millet

1 Hass avocado, pitted, peeled, and cubed

Your choice of pickled vegetables (in brine)

DIRECTIONS:

FOR THE VEGETABLES:

STEP 1. Preheat the oven to 400°F. Place the vegetables in a large bowl and toss them with salt and olive oil (and any other spices or seasoning) to taste.

STEP 2. Place the vegetables in a baking pan and roast for 25 minutes or until they begin to caramelize. Remove from the oven and set aside.

FOR THE DRESSING:

STEP 1. Soak the cashews for at least 30 minutes or overnight. Rinse the nuts and discard the water.

STEP 2. Place the rest of the dressing ingredients into a food processor and blend on high until creamy.

TO PREPARE THE BOWL:

STEP 1. Place the vegetables in a large serving bowl. Arrange the grains on top of the greens. Top with the avocado and pickled veggies.

STEP 2. Drizzle the dressing over the vegetables and grains. Sprinkle with the extra hemp seeds.

Cannellini Beans and Ancient Grains Soup

Cannellini beans are especially pleasing to the palate in the company of fresh herbs and ancient grains like amaranth and buckwheat. These ingredients together make for a hale and hearty soup filling enough to be a main dish. For a super-rich and creamy soup, puree all the soup rather than leaving half the beans whole.

INGREDIENTS ***(Serves 4)***

2 tablespoons extra-virgin olive oil

3 leeks, white parts only, sliced

3 garlic cloves, minced

½ cup amaranth

½ cup buckwheat

3 sprigs fresh rosemary, minced

10 fresh sage leaves, minced

1 bay leaf

1 tablespoon organic concentrated tomato paste

2 cups vegetable broth

2 cups cooked cannellini beans, rinsed and drained

1 pinch brewer's yeast per serving

Sea salt and pepper to taste

DIRECTIONS

STEP 1. Heat the olive oil in a large, heavy saucepan over medium heat. Add the leeks and cook, stirring frequently, until golden and soft, about 5 minutes. Add the garlic and cook for 1 minute more, then add the amaranth, buckwheat, rosemary, sage, bay leaf, and tomato paste and stir to combine.

STEP 2. Add the broth and bring to a boil. Reduce the heat to low, cover, and simmer for 30 minutes.

STEP 3. Remove from the heat and let cool slightly. Remove the bay leaf. Add 1 cup of the beans and use a handheld immersion blender or potato masher to puree the mixture in the pot until smooth. Stir in the remaining beans, sprinkle with the brewer's yeast, and season with salt and pepper to taste.

Chickpeas Tikka Masala

Chickpeas are a lovely, substantial source of vegetable protein that marries beautifully with the lively spices of garam masala, the flavors characteristic of India's famous tikka masala. Garam masala is a fragrant Indian spice blend that includes ground cumin, coriander, cinnamon, cardamom, and black pepper. Each of these spices contains numerous phytonutrients known to have powerful antioxidant properties and support digestion. They also have carminative effects (which means they'll get rid of bloating). Plus, while adding a touch of natural sweetness, cinnamon helps lower blood pressure and stabilize blood sugar levels.

If you can't find garam masala at your grocery store, you can easily make your own or order it online. I like to add turmeric to the mix for extra antioxidant protection and flavor.

This recipe is intended to be a vegetarian version of the more common chicken tikka masala. Chickpeas replace the chicken, and coconut milk replaces the usual yogurt and heavy cream. This is delicious alone or served over rice. Brown basmati rice works particularly well in this recipe.

INGREDIENTS ***(Serves 4)***

2 tablespoons extra-virgin coconut oil or organic ghee

1 red onion, finely diced

4 garlic cloves, finely minced

Pinch of sea salt

1 tablespoon garam masala

1 teaspoon ground turmeric

One 2-inch piece fresh ginger, grated

3 cups organic chickpeas, cooked, drained, and rinsed

1 (28-ounce) can organic diced tomatoes

1 cup full-fat coconut milk

1 tablespoon organic concentrated tomato paste

1 handful fresh cilantro leaves, coarsely chopped

DIRECTIONS:

STEP 1. Heat the oil in a large saucepan over medium heat. Add the onion, garlic, and salt and stir. Sauté until the onion is partly translucent and slightly browned around the edges, 3 to 4 minutes.

STEP 2. Stir in the garam masala, turmeric, and ginger and cook until very fragrant, 1 to 2 minutes.

STEP 3. Add the chickpeas, and the tomatoes and their juice and bring to a boil. Reduce the heat to low and simmer for 15 minutes. Stir in the coconut milk, tomato paste, and ghee or oil and return to a simmer for 5 minutes. Remove from the heat and stir in the cilantro.

Dad's Lemon Roasted Chicken

My Dad's Lemon Roasted Chicken is a discovery definitely worth trying out. In a tangy twist on a classic chicken recipe, a whole (organic free-range) chicken is rubbed inside and out with aromatic herbs, then baked with freshly squeezed lemon juice and extra-virgin olive oil for a light and flavorful Italian dish. Paired with roasted potatoes, this dish makes for a perfect meal. Additionally, this recipe is chockful of brain-essential nutrients, in particular all essential amino acids that our brains use to make neurotransmitters like dopamine and serotonin. Plus, sage and rosemary have long been praised for their memory-supportive effects and nerve-toning qualities.

INGREDIENTS *(Serves 6)*

1 whole organic free-range chicken (approximately 2 pounds)

6 garlic cloves

4 sprigs fresh rosemary

Small bunch of fresh sage

2 teaspoons extra-virgin olive oil

Juice of 1 lemon

2 teaspoons Himalayan pink salt

DIRECTIONS:

STEP 1. Preheat oven to 300°F.

STEP 2. Place the chicken in a roasting pan, breast side down. Scatter the garlic cloves around the pan, as well as inside the chicken. Stuff the chicken with rosemary and sage. Add olive oil to the pan and pour over the lemon juice. Sprinkle with the salt.

STEP 3. Place the pan in the oven and cook for 60 minutes, or until juices run clear.

Nonna's Dandelion Greens with Lemon Juice and Extra-Virgin Olive Oil

Dandelion greens double as both a delicious food and herbal medicine that anyone can find, grow, and put to good use. This flowering plant is rich in vitamins A and C, several B vitamins, iron, potassium, zinc, and fiber. Additionally, it contains a number of nutrients that feed the friendly bacteria in your gut. This is my *nonna*'s original recipe, which we used to enjoy almost every weekend.

INGREDIENTS *(Serves 4)*

1 pound organic dandelion greens

1 quart filtered water

2 tablespoons extra-virgin olive oil

Juice of 1 lemon

Pinch of sea salt

DIRECTIONS:

STEP 1. Rinse the greens and place them in a large saucepan. Cover with the water and bring to a boil over medium heat. Cook for 8 to 10 minutes or until the greens are tender but not mushy. Drain well.

STEP 2. Transfer to a serving bowl and drizzle with the olive oil and lemon juice and add the salt to taste.

Essential Vegetable Soup

This is the ultimate brain-healthy soup, featuring a wide range of super nutrients. Sweet peas are a good source of glutathione, our body's master antioxidant. Onions are healthy carbohydrates, rich in glucose and good at nourishing our good-gut microbes. Broccoli is a nutritional powerhouse, high in fiber and vitamins A, C, and B6, along with plenty of antioxidant phytonutrients. Edamame is a good source of lean vegetable protein that our brains rely on to function properly. Finally, brewer's yeast is an excellent source of brain-essential choline and vitamin B12. If that weren't enough, these flavors combine to make a knockout of a dish. Ideally, use only organic vegetables. I prefer to cook these al dente, as I like the consistency better than mushy vegetables and I believe it preserves the veggies' ability to better deliver their nutrients.

INGREDIENTS *(Serves 6)*

1 pound broccoli, finely chopped

1 cup finely chopped red cabbage

6 medium carrots, finely chopped

6 scallions, finely chopped, white tops only

4 stalks organic celery, finely chopped

4 garlic cloves, finely chopped (roasted garlic is even better!)

2 cups organic sweet peas (frozen are good, too)

1 cup organic edamame (shelled and frozen are good, too)

One 1-inch piece fresh ginger, grated

2 quarts vegetable broth (no added salt)

6 teaspoons brewer's yeast, 1 teaspoon per person

DIRECTIONS:

STEP 1. Place all the veggies in a large pot. Add the broth. Bring to a boil, cover and simmer for 20 minutes or until the veggies are tender.

STEP 2. Distribute the soup among bowls. Sprinkle 1 teaspoon of the brewer's yeast over each serving. Feel free to add brown rice for extra texture.

Go-to Green Salad

This salad is a nutritional dynamo and a main staple in our household. Typically, I make a large batch of this salad, which becomes our lunch for one meal and then makes another appearance as a side over the next several days. To make sure it stays fresh, I store the salad in airtight glass containers in the fridge. This recipe features all sorts of fragrant, verdant greens, combined with glucose-rich scallions, soothing baby fennel, and heart-healthy avocado and olives. In addition, the fresh sauerkraut and radishes provide probiotic benefits while adding extra crunch.

(A word about radishes. These vegetables are composed of indigestible carbs, which facilitate digestion and remove toxins. Additionally, radishes are a great source of anthocyanins, those same phytonutrients with antioxidant properties that give blueberries and cherries their beautiful color.)

INGREDIENTS ***(Serves 4)***

FOR THE SALAD:

1 cup chopped baby kale

1 cup chopped baby spinach

1 cup chopped mixed greens

4 scallions, white part only, sliced

½ sweet yellow onion, thinly sliced

4 or 5 radishes, thinly sliced (my favorite is the rapanello, a round, baby radish with fuchsia skin and white flesh; it has a very pungent, delicately bitter taste)

1 bulb baby fennel, thinly sliced

¼ cup Kalamata olives, pitted and chopped

½ cup fresh sauerkraut or pickled cabbage

1 ripe avocado, peeled, pitted, and cubed

½ cup fresh blackberries or blueberries (best when in season)

(continued)

FOR THE DRESSING:

1 tablespoon flaxseed oil

Juice of ½ lemon

1 tablespoon apple cider vinegar

DIRECTIONS:

STEP 1. Place all the veggies in a large bowl and mix.

STEP 2. In a blender, place all the ingredients for the dressing and blend on high until well combined. Pour the dressing over the salad and mix until coated. Feel free to top with your favorite seeds—hemp, sunflower, or pumpkin—raisins, date crumbles, and even hazelnuts. Serve immediately.

Grilled Sweet Potatoes with Fresh Spinach Salad and Yogurt-Tahini Sauce

If you want to add a little color to your meals, Okinawan sweet potatoes will do the trick. Also known as *beni imo*, this deep purple spud is actually part of the morning glory family, a vine coveted for its beautiful deep purple flowers. Rich in flavor and packed with nutritional benefits, the Okinawan sweet potato is thought to be one of the reasons the Okinawans are among the world's longest-living people, suffering far less from age-related ailments such as heart disease, cancer, diabetes, and Alzheimer's. If you don't have access to Okinawan sweet potatoes or Stokes sweet potatoes, you can replace them with regular yams.

This recipe features a delicious yogurt-tahini sauce, which adds a tangy kick to the mix. Not there by accident, sesame seeds are an especially excellent brain food. They were worth their weight in gold during the Middle Ages, and with good reason. Besides having antioxidant and anti-inflammatory properties, these tiny seeds are a great source of tryptophan, which the brain uses to make serotonin, the feel-good neurotransmitter.

INGREDIENTS ***(Serves 4)***

4 Okinawan sweet potatoes or organic Stokes sweet potatoes

3 tablespoons extra-virgin coconut oil

1 cup organic full-fat yogurt

2 tablespoons organic tahini

2 tablespoons organic maple syrup

4 cups baby spinach

2 tablespoons extra-virgin olive oil

Juice of ½ lemon

Salt and pepper

DIRECTIONS:

STEP 1. Preheat the grill or grill pan to high heat.

STEP 2. Cut the sweet potatoes lengthwise into ½-inch-thick slices. In a large pot, bring 3 cups of water to a boil over high heat. Blanch the sweet potatoes for 2 to 3 minutes. Let cool and pat dry.

STEP 3. Brush the grill with coconut oil. When the oil is sizzling, place the sweet potatoes in a single layer on the grill and cook until charred, about 5 minutes per side.

STEP 4. In a small bowl, whisk together the yogurt, tahini, and maple syrup. Set aside.

STEP 5. In a medium bowl place the spinach. Drizzle with the olive oil and lemon juice. Toss to coat.

STEP 6. Divide the spinach among four plates. Top with the sweet potatoes, season with salt and pepper to taste, and drizzle the yogurt-tahini dressing on top. Serve warm.

Grilled Salmon in Ginger Garlic Marinade

This is my go-to recipe whenever I don't want to do anything complicated but still want to serve a nice dinner to my family. It's really the quality of the ingredients that makes the difference in this dish. In particular, wild salmon is loaded with protein and the two blockbuster omega-3s, DHA and EPA—star supporters of our brains, nerves, and eye development. As the body can't make omega-3 fatty acids, the best way to obtain them is through the food we eat—and wild salmon is one of the best sources on the planet. This recipe boosts the flavor of the fish with a marinade that does the trick in just a few hours.

INGREDIENTS ***(Serves 2)***

4 tablespoons unrefined canola oil

One 1-inch piece fresh ginger, grated

3 cloves garlic, minced

Juice of ½ lemon

1 tablespoon organic maple syrup

2 tablespoons tamari (best if organic)

6 ounces wild Alaskan salmon fillet, patted dry

Sea salt and cayenne pepper

DIRECTIONS:

STEP 1. To make the marinade, combine 3 tablespoons of the oil, the ginger, garlic, lemon juice, maple syrup, and tamari in a zip-lock plastic bag. Shake well. Add the fish to the bag, reseal, and shake again until it's well coated. Place the bag in the fridge for 3 to 4 hours.

STEP 2. Preheat the grill or grill pan to high heat. Remove the fish from the bag and discard the marinade. Add salt and pepper to taste.

STEP 3. Brush the grill with the remaining oil. When the oil is sizzling, place the fish on the grill and cook until lightly charred, about 4 minutes per side.

Lentil Dal with Spinach

Lentils are a good source of fiber and complex carbs that will keep you fuller for longer while also feeding your brain with time-released glucose. This recipe includes the warming herbs cumin and cardamom, as well as turmeric, a staple of Indian cuisine. Turmeric contains curcumin, a strong clinical-trial-tested antioxidant long known for its anti-aging effects. Additionally, curcumin has the added benefit of boosting DHA in the brain and facilitating the conversion of ALA (the form of omega-3 found in plants) to DHA. Since ALA is normally ineffectively converted to DHA, adding turmeric to vegetarian dishes is particularly useful in getting the job done, not to mention being delicious.

INGREDIENTS ***(Serves 4)***

2 teaspoons extra-virgin olive or coconut oil

1 yellow onion, finely chopped

1 teaspoon ground cumin

¼ teaspoon ground cardamom

4 cloves garlic, finely chopped

2 tablespoons finely chopped ginger

2 cups red lentils, rinsed and drained

4 cups vegetable broth

1½ cups chopped fresh tomatoes with their juice

2 cups chopped spinach or Swiss chard

⅓ cup chopped fresh cilantro

1 teaspoon ground turmeric

(continued)

½ cup full-fat coconut milk

Sea salt

DIRECTIONS:

STEP 1. Heat the oil in a large pot over medium-high heat. Add the onions and cook until softened, about 5 minutes.

STEP 2. Add the cumin, cardamom, garlic, and ginger and cook, stirring often, until fragrant, about 2 minutes.

STEP 3. Add the lentils, broth, tomatoes and their juice, spinach, cilantro, turmeric, coconut milk, and salt to taste and bring to a boil. Reduce the heat to medium-low, cover, and simmer, stirring often, until the lentils are soft, about 15 minutes. Ladle into bowls and serve.

Pretzel-Encrusted Tilapia Fillet

My husband made this dish a while back and it's been a big favorite of ours ever since. If you don't consider yourself a fish lover, this is the recipe that will win you over. Just like classic fish sticks or a fish 'n' chips–style meal, this dish is delectably crunchy and popular, but this time the crunch comes from pretzels. If that weren't enough, it's super simple to make and provides all the things fish is lauded for, from omega-3s to complete protein, while minimizing cholesterol and saturated fat consumption at the same time. One 8-ounce fillet, cut in half, should be perfect for two people.

INGREDIENTS *(Serves 2)*

½ cup whole-wheat flour

Sea salt (only if you are using unsalted pretzels)

1 organic cage-free egg

1 cup pretzels (best if whole wheat or sprouted grain)

8 ounces tilapia fillet

2 tablespoons unsalted, organic grass-fed butter or extra-virgin coconut oil

Juice of ½ lemon

DIRECTIONS:

STEP 1. Place the flour on a large plate and season with the salt.

STEP 2. In a large, shallow bowl, beat the egg.

STEP 3. Using a food processor, crush the pretzels as finely as possible and transfer to a plate.

STEP 4. Dip the tilapia in the flour first, then in the egg, allowing the excess to drip off. Dredge in the pretzel crumbs to coat.

STEP 5. Heat the butter in a large pan over medium-high heat. Add the tilapia in a single layer. Cook, without moving the fillets, for about 3 minutes, until the crust is golden on the bottom. Flip over and cook another 3 minutes until the other side is golden and the fish is cooked through. Sprinkle lemon juice over fish. Serve immediately. This pairs nicely with a simple green salad.

SNACKS

Brain-Healthy Trail Mix

The story goes that the first trail mix was invented in the 1960s as a way to refuel the body while hiking or doing ongoing strenuous activity. Since it was relatively lightweight and portable, and also full of energy-dense ingredients like dried fruit, nuts, and chocolate, it was perfect for trailside noshing. Old-school trail mix contains peanuts and raisins, but we're going to combine all sorts of higher-quality dried fruit, nuts, and seeds to make a high-powered snack. Needless to say, this trail mix is packed with brain-essential nutrients most of all.

INGREDIENTS ***(Serves 12)***

½ cup raisins

½ cup date crumbles

¼ cup sunflower seeds

¼ cup pumpkin seeds

¼ cup Brazil nuts

¼ cup goji berries

¼ cup roasted hazelnuts

½ cup halved walnuts

½ cup unsweetened coconut flakes

¼ cup unsweetened cacao nibs

¼ cup pistachios, shelled

½ cup sliced almonds

¼ cup hemp hearts

½ cup unsweetened banana chips

DIRECTIONS:

Combine all the ingredients and stash the mix in an airtight container. It will keep for up to 2 weeks in the refrigerator.

Peanut Butter Power Bites

This is a healthy take on the no-bake cookie. So good *and* so good for you. Every time I make these power bites, I am begged for the recipe. They are a hit with both kids and adults, and always disappear quickly. For a real superfood treat, add 1 teaspoon spirulina powder and 1 teaspoon maca root powder to the mix. Spirulina will make the mixture turn green, but the essential amino acids it provides for your brain are worth a little color!

INGREDIENTS *(Serves 8)*

1 cup rolled oats (you can choose gluten-free if you prefer)

½ teaspoon ground cinnamon

7 or 8 Medjool dates

Dash of maple syrup

3 tablespoons smooth organic peanut butter

½ cup chopped peanuts

DIRECTIONS:

STEP 1. In a food processor, combine the oats and cinnamon. Blend until the oats reach a flourlike consistency. Add the dates and maple syrup. Blend until a paste forms.

STEP 2. Add the peanut butter and blend until well combined and doughy. Depending on your food processor, you might need to add a couple of tablespoons of warm water to reach the desired consistency.

STEP 3. Working with about 1 tablespoon of mixture at a time, roll into 12 balls. Roll each ball in the peanuts. Refrigerate for one hour, then serve.

SMOOTHIES

A smoothie is a convenient snack option and a smart way to deliver concentrated nutrition in a delicious, convenient treat that you can take anywhere! Just blend, pour, and enjoy.

While smoothies alone won't fix a poor diet, they are an easy way to incorporate more fruits and vegetables as well as specific brain-essential nutrients into your diet. Personally, I have learned to appreciate the therapeutic properties of whole-food smoothies made from fresh, organic fruit, vegetables, and nuts and seeds. I also mix in select natural supplements known to boost brain health and cognitive abilities, such as sage extract, ginkgo, and ginseng.

Below are my personal recommendations for each neuro-nutritional level: Beginner, Intermediate, and Advanced. Enjoy!

Nourishing Green Smoothie

INGREDIENTS ***(Serves 2)***

1 cup coconut water

½ cup goat's milk or almond milk

1 handful of raw almonds

1 tablespoon chia seeds

1 teaspoon flaxseeds

1 teaspoon açaí berry powder

1 teaspoon goji berries

1 tablespoon raw unsweetened cacao powder

1 teaspoon maca powder

1 tablespoon organic spirulina powder

Red Panax ginseng extract with royal jelly and bee pollen (5 cc) (optional)

Sage extract (organic certified, alcohol-free; 1 ml) (optional)

DIRECTIONS:

Combine all the ingredients in a high-speed blender. Mix for 1 minute. Enjoy.

Soothing Cacao Smoothie

INGREDIENTS *(Serves 2)*

1 tablespoon raw unsweetened cacao powder

1 tablespoon almond meal

1 tablespoon chia seeds

1 teaspoon goji berries

1 tablespoon organic aloe vera juice

¼ cup chocolate (or vanilla) vegan or whey protein powder

1 cup coconut water

1 cup full-fat coconut milk

Red Panax ginseng extract with royal jelly and bee pollen (10 cc) (optional)

DIRECTIONS:

Combine all the ingredients in a high-speed blender. Mix for 1 minute on high. Enjoy.

Spicy Berry Smoothie

INGREDIENTS *(Serves 2)*

1 tablespoon açaí berry powder (or ⅓ package frozen açaí berries)

Handful of frozen blueberries

1 teaspoon goji berries

One 1-inch piece fresh ginger

½ organic Red Delicious apple

1 tablespoon organic spirulina powder

Pinch of cayenne pepper

(continued)

½ teaspoon ground turmeric

1 tablespoon maple syrup

2 cups filtered water

1 (240 mg v-cap) ginkgo biloba (optional)

DIRECTIONS:

Place all the ingredients in a high-speed blender. Blend for 1 minute on high. Enjoy.

ACKNOWLEDGMENTS

The research that culminated in this book was a team effort. I want to express my deepest appreciation to the many colleagues and collaborators who clenched their teeth and soldiered on in the face of back-to-back deadlines, countless rejections, unyielding reviewers, last-minute changes, tracer failures, budget deficits, study audits, and so much more. It is their persistence and determination, as well as that of scientists all around the world, that allow us to finally and thoroughly explore the power of *prevention* of brain diseases like Alzheimer's.

I am indebted to the National Institute on Aging of the National Institutes of Health, the Alzheimer's Association, and several private foundations and generous benefactors for their continued involvement and support throughout the years. Without them, this research would just not have been possible.

A big thank-you to my girls, especially Kimberli, Lauren, Silvia, Bonnie, Amber, and Ramona, for always being there for me despite my being a phantom this past year; my late friend Kenneth Rich, MD, who taught me the true value and meaning of loving-kindness, and also introduced me to farmers' markets and health food stores; on the Italian front, my friends of a lifetime, especially Sonia, Gaia, Elena, Francesca, Foscarina, Valeria, Checco, Isa, Simone, and Franchina. I miss you all so much.

A special thank-you to my American *sorellina*, Susan Verrilli Dutilh. Not only did you introduce me to peanut butter and jelly at age five, but you were by my side every step and every page of this book, with endless patience and infinite kindness.

Heartfelt thanks to my editor, Caroline Sutton, and my literary agents, Katinka Matson and John Brockman, for giving me the opportunity to write this book in the first place. I am truly grateful for your tremendous support and outstanding expertise in transforming concepts and ideas into a tangible tool that can be put to work for better health and the greater good.

Last but not least, my family. *Mamma e papa'*, thank you for your unwavering love and support, even with an ocean in between—and for showing me the way around both the science laboratory and the kitchen. Huge thanks to *nonna* Marj, who patiently proofread this book several times.

Finally, *alle luci dei miei occhi*, Kevin and Lily, I love you more than anyone has ever loved anyone.

NOTES

Chapter 1: The Looming Brain Health Crisis

Page 3: According to the Centers for Disease Control: Zhaurova K. *Nature Education* 2008; 1:49.

Page 5: As the baby boomer generation ages: Barnes DE, Yaffe K. *Lancet Neurology* 2011; 10:819–828.

Page 5: the burden of all these disorders is reaching an alarming proportion: World Health Organization (WHO). www.who.int/mental_health/neurology/neurological_disorders_report_web.pdf.

Page 7: showing how it occurs gradually in the brain: Sperling RA et al. *Nature Reviews Neurology* 2013; 9:54–58.

Page 7: cognitive impairment is not a mere consequence of old age: Mosconi L et al. *Neurology* 2014; 82:752–760.

Page 7: that the brain changes leading to dementia can begin: Reiman EM et al. *Proceedings of the National Academy of Sciences USA* 2004; 101:284–289.

Page 8: While some neurons do continue to grow as we age: Aimone JB et al. *Nature Neuroscience* 2006; 9:723–727.

Page 8: the wear and tear that naturally occur as part of the aging process: Lazarov O et al. *Trends in Neuroscience* 2010; 33:569–579.

Page 10: our reserve will eventually be exhausted: Stern Y. *Lancet Neurology* 2012; 11:1006–1012.

Page 10: clinical trials have yielded mostly disappointing results: Mangialasche F et al. *Lancet Neurology* 2010; 9:702–716.

Page 11: ***less than 1 percent*** **of the population develops Alzheimer's:** Tanzi RE, Bertram L. *Neuron* 2001; 32:181–184.

Page 11: from the interplay of a multitude of genetic and lifestyle factors: Jimenez-Sanchez G et al. *Nature Genetics* 2001; 409:853–855.

Page 12: Research on identical twins is particularly enlightening: Herskind AM et al. *Human Genetics* 1996; 97:319–323.

Page 12: it was estimated that 70 percent of all cases of stroke: Willett WC. *Science* 2002; 296:695–698.

Page 12: addressing just a few of the risk factors for heart disease and diabetes: Norton S et al. *Lancet Neurology* 2014; 13:788–794.

Page 13: In addition to being toxic and depleting our soil: Davis DR et al. *Journal of the American College of Nutrition* 2004; 23:669–682.

Page 17: people who follow a Mediterranean diet: Mosconi L et al. *Journal of Prevention of Alzheimer's Disease* 2014; 1:23–32.

Page 17: regardless of whether or not they carry genetic risk factors for dementia: Mosconi L, McHugh PF. *Current Nutrition Reports* 2015; 4:126–135.

Chapter 2: Introducing the Human Brain, a Picky Eater

Page 22: It is made of a wall of flattened cells: Segal M. Blood-brain barrier. In Blakemore C, Jennett S, eds. *The Oxford Companion to the Body*. New York: Oxford University Press, 2001.

Page 24: research in evolutionary biology has shown: Leonard WR et al. *Annual Review of Nutrition* 2007; 27:311–27.

Page 26: Africa has long been agreed upon as the cradle of humanity: Stringer C. *Nature* 2003; 423:692–693.

Page 26: Grasses, seeds and sedges, fruits, roots, bulbs, tubers: Teaford MF, Ungar PS. *Proceedings of the National Academy of Sciences USA* 2000; 97:13506–13511.

Page 28: that spurred the brain's expansion by providing energy-dense "brain food": Cunnane SC et al. *Nutrition and Health* 1993; 9:219–235.

Page 28: these foods required little skill to fetch and consume: Broadhurst CL et al. *British Journal of Nutrition* 1998; 79:3–21.

Page 28: early humans participated in "confrontational scavenging": Joordens JC et al. *Nature* 2015; 518:228–231.

Page 29: This higher-quality diet further increased our ancestors' fat consumption: Ungar PS. *Journal of Human Evolution* 2004; 46:605–622.

Page 29: When meat and fruit were scarce: Eaton SB, Konner M. *New England Journal of Medicine* 1985; 312:283–289.

Page 30: no more than 25 to 35 percent: Cordain L et al. *American Journal of Clinical Nutrition* 2000; 1589–1592.

Page 30: Several research teams have documented: Henry AG, et al. *Nature* 2012; 487:90–93.

Page 30: how ancient grains like oats and wild wheat: Cerling TE et al. *Proceedings of the National Academy of Sciences USA* 2013; 110:10501–10506.

Page 30: It could very well be the development of habitual cooking: Wrangham R et al. *Current Anthropology* 1999; 40:567–594.

Page 31: But thanks to the increased access to animal foods and their own cooking skills: Aiello LC, Wheeler P. *Current Anthropology* 1995; 36:199–221.

Page 32: our ancestors' diet couldn't be more different from ours: Eaton SB, Eaton SB, III. *European Journal of Nutrition* 2000; 39:67–70.

Page 32: our fat consumption is relegated to processed baked goods: U.S. Department of Agriculture, Agricultural Research Service, 1997. https://www.ncbi.nlm.nih.gov/pmc/articles/PMC1929441/pdf/pubhealthreporig00112-0059.pdf.

Page 33: the disease-causing genes known to scientists thus far: Wellcome Trust Case Consortium. *Nature* 2007; 447:661–678.

Chapter 3: The Water of Life

Page 36: our bodies are made of a fair amount of water: McIlwain H, Bachelard HS. *Biochemistry and the Central Nervous System* (5th edition). Edinburgh: Churchill Livingstone, 1985.

Page 37: first living creatures were born in the depths of the oceans: Maher KA, Stevenson DJ. *Nature* 1988; 331:612–614.

Page 37: brain cells require a delicate balance of water: Amiry-Moghaddam M, Ottersen OP. *Nature Reviews Neuroscience* 2003; 4:991–1001.

Page 38: causing a number of issues like fatigue: Popkin BM et al. *Nutrition Reviews* 2010; 68:439–458.

Page 38: 43 percent of adult Americans report: Goodman AB et al. *Prevention of Chronic Disease* 2013; 10:E51.

Page 38: when we are dehydrated: Streitburger DP et al. *PLoS One* 2012; 7:e44195.

Page 38–39: If you need more of an incentive: Benefer MD et al. *European Journal of Nutrition* 2013; 52:617–624.

Page 39: Researchers in the UK ran an experiment: Edmonds CJ et al. *Frontiers Human Neuroscience* 2013; 7:363.

Page 39: making older people more vulnerable: Farrell MJ et al. *Proceedings of the National Academy of Sciences USA* 2008; 105:382–387.

Page 40: carbonated soft drinks are the most-consumed beverages: LaComb RP et al. *Food Surveys Research Group Dietary Data Brief No. 6.* August 2011.

Page 41: plain water that is high in minerals: Haas EM. *Staying Healthy with Nutrition: The Complete Guide to Diet & Nutritional Medicine.* Berkeley, CA: Celestial Arts, 1992.

Chapter 4: The Skinny on Brain Fat

Page 45: Most people are aware that the human brain is rich in fat: McIlwain H, Bachelard HS. *Biochemistry and the Central Nervous System* (5th edition). Edinbugh: Churchill Livingstone, 1985.

Page 46: fat accounts for less than half the brain's weight: Brady S et al. *Basic Neurochemistry: Principles of Molecular, Cellular, and Medical Neurobiology* (8th edition). Amsterdam: Elsevier, Academic Press, 2012.

Page 47: account for as much as 70 percent of all the fat found in the brain: O'Brien JS, Sampson EL. *Journal of Lipid Research* 1965; 6:545–551.

Page 51: the brain is able to make as much saturated fat: Sastry PS. *Progress in Lipids Research* 1985; 24:69–176.

Page 51: the brain might take up a little bit: Pardridge WM, Mietus LJ. *Journal of Neurochemistry* 1980; 34:463–466.

Page 51: Their tail has to be fairly short: Mitchell RW et al. *Journal of Neurochemistry* 2011; 117:735–746.

Page 52: is largely "homemade" on the brain's premises: Sastry PS. *Progress in Lipids Research* 1985; 24:69–176

Page 52: PUFAs are the only kinds of fat the brain cannot make: Bachelard HS. *Brain Biochemistry* (2nd edition). London: Chapman and Hall, 1981.

Page 52: The brain is specifically designed to collect these fats: Edmond J. *Journal of Molecular Neuroscience* 2001; 16:181–193.

Page 52: omega-3s and omega-6s, are the best known: Williams CM, Burdge G. *Proceedings of the Nutrition Society* 2006; 65:42–50.

Page 53: the balance of these two PUFAs: Morris MC, Tangney CC. *Neurobiology of Aging* 2014; 35 Suppl 2:59–64.

Page 53: a ratio of two-to-one: Simopoulos AP. *American Journal of Clinical Nutrition* 1991; 54:438–463.

Page 53: estimates that Americans consume *twenty or thirty times* more: Kris-Etherton PM et al. *American Journal of Clinical Nutrition* 2000; 71:S179–188.

Page 53: This is ten times the amount: Food and Nutrition Board of the Institute of Medicine of the National Academies. *Dietary Reference Intakes for Energy, Carbohydrate, Fiber, Fat, Fatty Acids, Cholesterol, Protein, and Amino Acids (2002/2005).*

Page 55: people who consumed low quantities of omega-3s: Morris MC et al. *Archives of Neurology* 2003; 60:194–200.

Page 56: those who didn't consume enough omega-3s: Pottala JV et al. *Neurology* 2015; 82:435–442.

Page 56: whereas those who consumed 6 grams or more: Tan ZS et al. *Neurology* 2012; 78:658–664.

Page 56: clinical trials have still failed to show significant changes: Fotuhi M et al. *Nature Clinical Practice Neurology* 2009; 5:140–152.

Page 56: natural sources like fish *rather than from supplements*: Morris MC, Tangney CC. *Neurobiology of Aging* 2014; 35 Suppl 2: S59–S64.

Page 58: eat foods rich in phospholipids: Fernstrom MH. *Nutritional Pharmacology*. New York: Liss AR Inc., 1981.

Page 60: people who consumed at least 24 grams of these fats a day: Morris MC et al. *Archives of Neurology* 2003; 60:194–200.

Page 61: too much saturated fat can increase the risk of heart disease: Djoussé L, Gaziano JM. *Current Atherosclerosis Reports* 2009; 1:418–422.

Page 61: those who consistently ate the most saturated fat: Morris MC et al. *Archives of Neurology* 2003; 60:194–200.

Page 61: those who ate half that amount (13 grams per day): Okereke OI et al. *Annals of Neurology* 2012; 72:124–134.

Page 63: trans fats and increased risk of cognitive decline: Barnard ND et al. *Neurobiology of Aging* 2014; 35:65–73S.

Page 63: people who consumed 2 or more grams of trans fats a day: Morris MC et al. *Archives of Neurology* 2003; 60:194–200.

Page 63: This increases our risk of cardiovascular disease: Mensink RP, Katan MB. *New England Journal of Medicine* 1990; 323:439–445.

Page 64: Due to some latitude in current regulations: Food and Drug Administration (11 July 2003). "FDA food labeling: trans fatty acids in nutrition labeling; consumer research to consider nutrient content and health claims and possible footnote or disclosure statements," p. 41059.

Page 66: The brain continues to make cholesterol: Orth M, Bellosta S. *Cholesterol* 2012; 2012:292–298.

Page 66: the brain completely seals it away: Di Paolo G, Kim TW. *Nature Reviews Neuroscience* 2011; 12:284–296.

Page 67: those with high cholesterol in midlife: Kivipelto M et al. *Annals of Internal Medicine* 2002; 137:149–155.

Page 67: A cholesterol level of 220 mg/dL: Solomon A et al. *Dementia Geriatric Cognitive Disorders* 2009; 28:75–80.

Page 68: only 25 percent or so is derived from the diet: Kanter M et al. *Advances in Nutrition* 2012; 3:711–717.

Page 68: more than consuming cholesterol itself: Djoussé L, Gaziano JM. *Current Atherosclerosis Reports* 2009; 1:418–422

Page 68: no association between eating eggs and the risk of heart disease: Orth M, Bellosta S. *Cholesterol* 2012; 12:292–298.

Page 69: a wide range of responses: Orth M, Bellosta S. *Cholesterol* 2012; 12:292–298

Page 69: PUFAs are the heart's main source of fuel: Berg JM et al. *Biochemistry* (5th edition). New York: W H Freeman, 2002.

Page 70: the brain might do well to take up: Mitchell RW et al. *Journal of Neurochemistry* 2011; 117:735–746.

Chapter 5: The Benefits of Protein

Page 72: When you eat a meal that contains protein: Laterra J et al. Blood-Brain Barrier. In Siegel GJ, Agranoff BW, Albers RW et al., eds. *Basic Neurochemistry: Molecular, Cellular and Medical Aspects* (6th edition). Philadelphia: Lippincott-Raven, 1999.

Page 73: composed of over 80 billion brain cells: Kandel ER et al. *Principles of Neural Science* (5th Edition). New York: McGraw-Hill, 2012.

Page 75: Less well-known is that its depletion: McEntee WJ, Crook TH. *Psychopharmacology* 1991; 103:143–149.

Page 75: the production of serotonin in the brain: Wurtman RJ, Fernstrom JD. *American Journal of Clinical Nutrition* 1975; 28:638–647.

Page 75: the average adult, man or woman, needs 5 mg of tryptophan: Food and Nutrition Board of the Institute of Medicine of the National Academies. *Dietary Reference Intakes for Energy, Carbohydrate, Fiber, Fat, Fatty Acids, Cholesterol, Protein, and Amino Acids (2002/2005).*

Page 77: eating carbohydrates with or immediately after tryptophan-rich foods: Wurtman RJ, Fernstrom JD. *American Journal of Clinical Nutrition* 1975; 28:638–647.

Page 78: Dopamine abnormalities are involved in several medical conditions: Calabresi P et al. *Trends in Neuroscience* 2000; 23:57–63S.

Page 78: Tyrosine needs to be produced from another amino acid: Fernstrom JD, Fernstrom MH. *Journal of Nutrition* 2007; 137:1539S–1547S.

Page 79: the recommended dose of phenylalanine and tyrosine: Food and Nutrition Board of the Institute of Medicine of the National Academies. *Dietary Reference Intakes for Energy, Carbohydrate, Fiber, Fat, Fatty Acids, Cholesterol, Protein, and Amino Acids (2002/2005).*

Page 81: It is widely believed that this process: Kalia LV et al. *Lancet Neurology* 200; 7:742–755.

Page 82: Glutamate is formed when the brain breaks down glucose: Shen J et al. *Proceedings of the National Academy of Sciences USA* 1999; 96:8235–8240.

Chapter 6: Carbs, Sugars, and More Sweet Things

Page 83: This awe-inspiring process requires: Du F et al. *Proceedings of the National Academy of Sciences USA* 2008; 105:6409–6414.

Page 84: the brain relies exclusively on a sugar: Sokoloff L. *Journal of Neurochemistry* 1977; 29:13–26.

Page 85: "sugar gates" are present in the blood-brain barrier: Sokoloff L. *Journal of Neurochemistry* 1977; 29:13–26.

Page 86: It still requires no less than 30 percent of its energy from glucose: Sokoloff L. *Annals Reviews Medicine* 1973; 24:271–280.

Page 88: the brain burns an average of 32 micromoles of glucose: Sokoloff L. *Journal of Neurochemistry* 1977; 29:13–26.

Page 90: 6 to 8 percent of all dementia cases are attributed to type 2 diabetes: Sims-Robinson C et al. *Nature Reviews Neurology* 2010; 6:551–559.

Page 90: stroke account for another 25 percent of patients: Morris MS. *Lancet Neurology* 2003; 2:425–428.

Page 90: insulin resistance can lead to brain inflammation: Biessels GJ, Reagan LP. *Nature Reviews Neuroscience* 2015; 16:660–671.

Page 90: compare high blood sugar with the possibility of poor cognitive outcomes: Crane PK et al. *New England Journal of Medicine* 2013; 369:540–548.

Page 91: high sugar levels exhibit not only decreased memory performance: Convit A et al. *Proceedings of the National Academy of Sciences USA* 2003; 100:2019–2022.

Page 91: This correlation was also found in participants without a trace of diabetes: Tiehuis AM et al. SMART Study Group. *Diabetes Care* 2014; 37:2515–2521.

Page 93: (such as the United States) top the list of low-fiber eaters: Ferlay J et al. GLOBOCAN 2012 v1.1, Cancer Incidence and Mortality Worldwide: IARC CancerBase No. 1. globocan.iarc.fr.

Page 93: "treats" still possess an overall low glycemic load: Willett W et al. *American Journal of Clinical Nutrition* 2002; 76:274–280S.

Chapter 7: Making Sense of Vitamins and Minerals

Page 96: When you eat fresh vegetables or fruits: Spector R. *Journal of Neurochemistry* 1989; 53:1667–1674.

Page 97: acetylcholine is limited by how much choline is reaching the brain: Wurtman RJ. *Trends in Neuroscience* 1992; 15:117–122.

Page 98: 90 percent of the American population is deficient in choline: Jensen HH et al. *The FASEB Journal* 2007; 21:lb219.

Page 98: According to current dietary guidelines: Food and Nutrition Board of the Institute of Medicine of the National Academies. *Dietary Reference Intakes for Thiamin, Riboflavin, Niacin, Vitamin B6, Folate, Vitamin B12, Pantothenic Acid, Biotin, and Choline* (1998).

Page 101: high homocysteine (*hyperhomocysteinaemia*) is a strong risk factor for stroke: Morris MS. *Lancet Neurology* 2003; 2:425–428.

Page 101: the risk of developing dementia was nearly doubled: Seshadri S et al. *New England Journal of Medicine* 2002; 346:476–483.

Page 102: those whose diets were rich in folate: Luchsinger JA et al. *Archives of Neurology* 2007; 64:86–92.

Page 102: those who had low B12 levels: Tangney CC et al. *Neurology* 2009; 72:361–367.

Page 102: Recent randomized, double-blind, placebo-controlled trials: Douaud G et al. *Proceedings of the National Academy of Sciences USA* 2013; 110:9523–9528.

Page 103: the treatment's success was also related to the patients' consumption of omega-3 PUFAs: Jernerén F et al. *American Journal of Clinical Nutrition* 2015; 102:215–221.

Page 104: the brain is the one that suffers most from oxidative stress: Jenner P. *Lancet* 1994; 344:796–798.

Page 105: 11 IU (16 mg) of vitamin E per day had a 67 percent lower risk of developing dementia: Morris MC et al. *JAMA* 2002; 287:3230–3237.

Page 105: vitamins C and E had an even lower risk: Engelhart MJ et al. *Journal of the American Medical Association* 2002; 287:3223–3229.

Page 105: there is consensus that regular consumption of vitamins C and E: Maden M. *Nature Reviews Neuroscience* 2007; 8:755–765.

Page 105: reduces the speed at which our brain cells age: Meydani M. *Lancet* 1995; 345:170–175.

Page 105: Vitamin E was the only one that showed potential: Dysken MW et al. *Journal of the American Medical Association* 2014; 311:33–44.

Page 106: This seems to reduce oxidative stress and inflammation: Liu M et al. *American Journal of Clinical Nutrition* 2003; 77:700–706.

Page 107: each essential in keeping our brains healthy: World Health Organization (WHO). *Neurological disorders associated with malnutrition.* www.who.int/mental_health/neurology/chapter_3_b_neuro_disorders_public_h_challenges.pdf.

Page 108: aluminum is toxic to brain cells: Bondy SC. *Neurotoxicology* 2016; 52:222–229.

Page 109: overconsuming iron, zinc, and copper might contribute to cognitive problems: Doraiswamy PM, Finefrock AE. *Lancet Neurology* 2004; 3:431–434.

Page 109: the copper we ingest just by eating the typical modern diet: Singh I et al. *Proceedings of the National Academy of Sciences USA* 2013; 110:14771–14776.

Page 109: people whose diets are high in copper, saturated fat, and trans fat: Morris MC et al. *Archives of Neurology* 2006; 63:1085–1088.

Chapter 8: Food Is Information

Page 112: a genetic mutation occurred: Eiberg H et al. *Human Genetics* 2008; 123:177–187.

Page 112: affecting less than 1 percent of the population: Tanzi RE, Bertram L. *Neuron* 2001; 32:181–184.

Page 112: a large part of which involve brain function: Sachidanandam R et al. *Nature* 2001; 409:928–933.

Page 113: to activate or silence your genes: Jirtle RL, Skinner MK. *Nature Reviews Genetics* 2007; 8:253–262.

Page 114: dietary nutrients have the ability to influence: Dauncey MJ. *Proceedings of the Nutrition Society* 2012; 71:581–591.

Page 114: *nutrigenomics,* which aims at revealing: Muller M, Kersten S. *Nature Reviews Genetics* 2003; 4:315–322.

Page 115: your cells are coded with detectors: Ibid.

Page 115–16: each human being has a unique biochemistry: Williams RJ. *Biochemical Individuality: The Basis for the Genetotrophic Concept.* New York: Wiley & Sons, 1956.

Page 116: many human genes have a heightened sensitivity to diet: Scriver CR. *American Journal of Clinical Nutrition* 1988; 48:1505–1509.

Page 116: keeping the lactase gene turned on: Tishkoff SA et al. *Nature Genetics* 2007; 39: 31–40.

Page 117: an adult human harbors nearly 100 trillion bacteria: Qin J, Li R, Raes J, Arumugam M et al. *Nature* 2010; 464:9–65.

Page 117: bacterial cells outnumber human cells: Turnbaugh PJ et al. *Nature* 2007; 449:804–810.

Page 117: the human genome (aka our DNA) is extremely small: Venter JC et al. *Science* 2001; 291:1304–1351.

Page 118: our gut microbes are major players in our overall health: Collins SM et al. *Nature Reviews Microbiology* 2012; 10:735–742.

Page 118: they can directly alter the function of the blood-brain barrier: Braniste V et al. *Science Translational Medicine* 2014; 6:263.

Page 118: "leaky gut" can occur: Fasano A et al. *Lancet* 2000; 355:1518–1519.

Page 119: This initial research has triggered tremendous interest: Mayer EA et al. *The Journal of Neuroscience* 2014; 34:15490–15496.

Page 120: animals genetically engineered to be *without a microbiome*: Sudo N et al. *Journal of Physiology* 2004; 558:263–275.

Page 120: directly increased production of GABA: Bravo JA et al. *Proceedings of the National Academy of Sciences USA* 2011; 108:16050–16055.

Page 120: a connection with issues present in the child's microbiome: Cryan JF, Dinan TG. *Nature Reviews Neuroscience* 2012; 13:701–712.

Page 121: It made them less anxious: Hsiao EY et al. *Cell* 2013; 155:1451–1463.

Page 121: eating probiotic foods like yogurt would elicit: Tillisch K et al. *Gastroenterology* 2013; 144:1394–1401.

Page 121: it might be a preeminent factor: Claesson MJ et al. *Nature* 2012; 488: 178–184.

Page 121–22: diets are low in fiber but high in animal fat: Ibid.

Page 125: According to a recent study by the U.S. Food and Drug Administration: United States Food and Drug Administration (FDA). Reports and Data. The 2012–2013 Integrated NARMS Report. www.fda.gov/downloads/AnimalVeterinary/SafetyHealth /AntimicrobialResistance/NationalAntimicrobialResistanceMonitoringSystem/ UCM453398.pdf.

Page 126: processed foods often contain *emulsifiers*: Chassaing B et al. *Nature* 2015; 519:92–96.

Page 127: Similar reactions are sometimes observed: Biesiekierski JR et al. *American Journal of Gastroenterology* 2011; 106:508–514.

Chapter 9: The World's Best Brain Diets

Page 131: The first of these longevity hotspots: Poulain M et al. *Experimental Gerontology* 2004; 39:1423–1429.

Page 133: Their beloved olive oil: Buettner D. *The Island Where People Forget to Die. New York Times Magazine*, 2012.

Page 133: Some classic staple foods: Willcox BJ et al. *Annals of the New York Academy of Science* 2007; 1114:434–55.

Page 134: Black beans, white rice, yams, and eggs: Rosero-Bixby L. *Demography* 2008; 45:673–691.

Page 134: It comes as little or no surprise: Fraser GE, Shavlik DJ. *Archives of Internal Medicine* 2001; 161:1645–1652.

Page 136: this oil contains heart-healthy monounsaturated fat: Owen RW et al. *Lancet Oncology* 2000; 1:107–112.

Page 136: clinical trials show that if we regularly consume extra-virgin olive oil: Vallas-Pedret C et al. *JAMA Internal Medicine* 2015; 175:1094–1103.

Page 136: Red wine is another main staple: Corder R et al. *Nature* 2006; 444:566.

Page 137: those who followed the diet had overall healthier brains: Mosconi L et al. *Journal of Prevention of Alzheimer's Disease* 2014; 1:23–32.

Page 137: their brains appear to be a good five years older: Gu Y et al. *Neurology* 2015; 85:1744–1751.

Page 137: Not only were these beleaguered brains shrinking: Matthews DC et al. *Advances in Molecular Imaging* 2014; 4:43–57.

Page 138: those who followed the Mediterranean diet during middle age: Samieri C et al. *Annals of Internal Medicine* 2013; 159:584–591.

Page 138: A new diet known as the MIND diet: Morris MC et al. *Alzheimer's & Dementia* 2015; 11:1007–1014.

Page 139–40: administering 240 mg/day of ginkgo extract for about six months: Tan MS et al. *Journal of Alzheimer's Disease* 2015; 43:589–603.

Page 140: Panax ginseng might be helpful in improving: Lee MS et al. *Journal of Alzheimer's Disease* 2009; 18:339–344.

Page 140: India has a spectacularly low incidence: Chandra V et al. *Neurology* 2001; 57: 985–989.

Page 140: mice that were fed curcumin: Lim GP et al. *Journal of Neuroscience* 2001; 21:8370–8377.

Page 140: only a few clinical trials of curcumin: Brondino N et al. *Scientific World Journal* 2014; 2014:174282.

Page 141: The antioxidant diet: Mattson MP, Magnus T. *Nature Reviews Neuroscience* 2006; 7:278–294.

Page 142: This in turn accelerates brain aging: Cai W et al. *Proceedings of the National Academy of Sciences USA* 2014; 111:4940–4945.

Page 142: Animal-derived foods high in fat: Uribarri J et al. *Journal of the American Dietary Association* 2010; 110:911–916.

Page 142: carbohydrate-rich foods contain relatively few AGEs: Ibid.

Page 143: caloric restriction, or dramatically reducing your calories: Mattson MP, Wan R. *Journal of Nutritional Biochemistry* 2005; 16:129–137.

Page 143: It also reduces inflammation: Ibid.

Page 143: caloric restriction does indeed lower the risk of memory loss: Witte AV et al. *Proceedings of the National Academy of Sciences USA* 2009; 106:1255–1260.

Page 144: laboratory animals by up to 30 percent: Longo VD, Mattson MP. *Cell Metabolism* 2014; 19:181–192.

Page 144: In a recent study: Harvie MN et al. *International Journal of Obesity* 2011; 35:714–727.

Page 145: Recent data suggests that the keto diet: Maalouf M et al. *Brain Research Reviews* 2009; 59:293–315.

Page 145: Although clinical trials have been scarce: Vanitallie TB et al. *Neurology* 2005; 64:728–730.

Page 145: Similarly, patients with Alzheimer's: Reger MA et al. *Neurobiology of Aging* 2004; 25:311–314.

Page 146: These greens come with an arsenal: Trichopoulou A. *Public Health Nutrition* 2004; 7:943–947.

Page 146: Many research studies have shown: Joseph J et al. *Journal of Neuroscience* 2009; 29:12795–12801.

Page 147: consumption of cocoa drinks with a high flavonoid content: Mastroiacovo D et al. *American Journal of Clinical Nutrition* 2015; 101:538–548.

Page 147: people who drink coffee daily in midlife: Eskelinen MH et al. *Journal of Alzheimer's Disease* 2009; 16:85–91.

Page 147: Red wine is a great source of *resveratrol*: Price NL et al. *Cell Metabolism* 2012; 15:675–690.

Page 147: clinical trials have so far failed: Wightman EL et al. *British Journal of Nutrition* 2015; 114:1427–1437.

Page 148: Green tea contains twice the amount of antioxidants: Pellegrini N et al. *Journal of Nutrition* 2003; 133:2812–2819.

Page 148: Green tea is also quite rich in a special flavonoid: Hyung SJ et al. *Proceedings of the National Academy of Sciences USA* 2013; 110:3743–3748.

Page 148: The result is improved cognitive function: Willis LM et al. *British Journal of Nutrition* 2009; 101:1140–1144.

Page 148: older people who consumed fish regularly: Kalmijn S et al. *Neurology* 2004; 62:275–280.

Chapter 10: It's Not All About Food

Page 152: The physically fit elderly typically perform better: Van Praag H. *Trends in Neuroscience* 2009; 32:283–290.

Page 152: exercise promotes heart health: Hillman CH et al. *Nature Reviews Neuroscience* 2008; 9:58–65.

Page 153: The more you work out: Cotman CW et al. *Trends in Neuroscience* 2007; 30:464–472.

Page 153: particularly effective at dissolving Alzheimer's plaques: Gleeson M et al. *Nature Reviews Immunology* 2011; 11:607–615.

Page 154: those who engaged in activities such as walking: Scarmeas N et al. *Journal of the American Medical Association* 2009; 302:627–637.

Page 154: even those who engaged in *light* physical activity: Ibid.

Page 154: much more pronounced in the sedentary elderly: Okonkwo OC et al. *Neurology* 2014; 83:1753–1760.

Page 155: a sedentary life is harmful to your brain: Matthews DC et al. *Advances in Molecular Imaging* 2014; 4:43–57.

Page 155: a study of 120 sedentary adults: Erickson KI et al. *Annals of Neurology* 2010; 68:311–318.

Page 157: This produced some remarkable results: McCay C et al. *Bulletin of the New York Academy of Medicine* 1956; 32:91–101.

Page 157: when older mice were given blood from their younger counterparts: Villeda SA et al. *Nature Medicine* 2014; 20:659–663.

Page 157: as people get older, their stem cells: Conboy IM et al. *Nature* 2005; 433:760–764.

Page 158: These blood proteins: Sinha M et al. *Science* 2014; 344:649–652.

Page 158: Several nutrients are thought to enhance: Mundy GR. *American Journal of Clinical Nutrition* 2006; 83:427–430S.

Page 159: Cardiovascular disease is a major risk: Kalaria RN et al. *Lancet Neurology* 2008; 7:812–826.

Page 159: The prescription is simple: 2013 ACC/AHA Guideline on the Assessment of Cardiovascular Risk: A Report of the American College of Cardiology/American Heart Association Task Force on Practice Guidelines. *Journal of the American College of Cardiology* 2014; 63:2935–2959.

Page 160: people who retire at an early age have an increased risk of developing dementia: Dufouil C et al. *European Journal of Epidemiology* 2014; 29:353–361.

Page 160: those who regularly engaged in intellectual activity: Verghese J et al. *Neurology* 2006; 66:821–827.

Page 161: lifelong participation in such activities: Landau SM et al. *Archives of Neurology* 2012; 69:623–629.

Page 161: call to arms against the brain-training industry: Max Planck Institute for Human Development and Stanford Center on Longevity. longevity3.stanford.edu /blog/2014/10/15/the-consensus-on-the-brain-training-industry-from-the-scientific -community.

Page 161: participation in a brain-training program: Willis SL et al. ACTIVE Study Group. *Journal of the American Medical Association* 2006; 296:2805–2814.

Page 162: this sort of cognitive training is only modestly effective: Lampit A et al. *PLoS Medicine* 2014; 11:e10001756.

Page 162: playing board games as the intellectual activity: Dartigues JF et al. *British Medical Journal* 2013; 3:e002998.

Page 163: elderly with stronger social networks: Holt-Lunstad J et al. *PLoS Medicine* 2010; 7:e1000316.

Page 163: having a family you love is enough: Fratiglioni L et al. *Lancet* 2000; 355:1315–1319.

Page 164: sleeping is crucial for memory consolidation: Stickgold R et al. *Science* 2001; 294:1052–1057.

Page 164–65: the brain's unique waste-removal technique: Iliff JJ et al. *Journal of Clinical Investigations* 2013; 123:1299–1309.

Page 165: brain clearing becomes ten times more active during sleep: Xie L et al. *Science* 2013; 342:373–377.

Page 165: older adults who slept less than five hours: Spira AP et al. *JAMA Neurology* 2013; 70:1537–1543.

Page 167: relatively easy lifestyle-based strategies to fight dementia: Ngandu T et al. *Lancet* 2015; 385:2255–2263.

Chapter 11: A Holistic Approach to Brain Health

Page 173: those who have mastered the secrets: Willcox BJ, et al. *Annals of the New York Academy of Science* 2007; 1114:434–55.

Page 174: the most popular vegetable in America: United States Department of Agriculture. ERS Food Availability (Per Capita) Data System (FADS), 2015. www.ers.usda.gov/data-products/food-availability-per-capita-data-system.

Page 182: These sweeteners have come under scrutiny: Mitka M. *Journal of the Medical American Association* 2016; 315:1440–1441.

Page 182: natural sweeteners come with an added bonus: Phillips KM, et al. *Journal of the American Dietetic Association* 2009; 109:64–71.

Page 187: than boiled or filtered coffee: Yashin A et al. *Antioxidants* 2013; 2:230–245.

Page 188: a much higher content of it: Carlsen MH et al; *Nutrition Journal* 2010; 9:3–10.

Page 188: organic pomegranate juice is almost as rich: Ibid.

Page 189: the combination of several nutrients: Berti V et al. *Journal of Nutrition Health and Aging* 2015; 19:413–423.

Page 189: particularly effective at protecting memory: Mosconi L et al. *British Medical Journal (Open Access)* 2014; 4:e004850.

Page 189–90: people who routinely consume these nutrients together: Berti V et al. *Journal of Nutrition Health and Aging* 2015; 19:413–423.

Page 190: exhibit more pronounced brain shrinkage: Bowman GL et al. *Neurology* 2012; 78:241–249.

Chapter 12: Be Mindful of Quality Over Quantity

Page 194: Soy products in the United States today: United States Department of Agriculture (USDA). www.ers.usda.gov/data-products/adoption-of-genetically-engineered-crops-in-the-us/recent-trends-in-ge-adoption.aspx.

Page 195: Soy is added to as many as twelve thousand food products: Anderson JW et al. *New England Journal of Medicine* 1995; 333:276–282.

Page 203: once you change your diet: Koeth RA et al. *Nature Medicine* 2013; 19:576–585

Page 203: Food processing, on the other hand, is a major problem: Stuckler D, Nestle M. *PLoS Medicine* 2012; 9:e1001242.

Page 204: significant declines in the amounts of vitamins: Davis DR et al. *Journal of the American College of Nutrition* 2004; 23:669–682.

Page 210: The recommended dose is 240 mg/day of ginkgo extract: Tan MS et al. *Journal of Alzheimer's Disease* 2015; 43:589–603.

Page 210: The recommended dose is 4 grams/day of red Panax ginseng powder: Lee MS et al. *Journal of Alzheimer's Disease* 2009; 18:339–344.

Chapter 13: A Typical Brain-Healthy Week

Page 225: This simple practice has been shown to reduce adipose body fat: Chaix A et al. *Cell Metabolism* 2014; 20:991–1005.

Page 226: people who engage in regular physical activity have a healthier microbiome: Clarke SF et al. *Gut* 2014; 63:1913–1920.

Chapter 15: The Three Levels of Neuro-Nutrition Care

Page 255: Red Delicious and Gala apples have the highest antioxidant content: Pellegrini N et al. *Journal of Nutrition* 2003; 133:2812–2819.

Page 266: wild greens were found to have ten times as many antioxidants as red wine: Trichopoulou A. *Public Health Nutrition* 2004; 7:943–947.

Page 266: people who eat one to two servings of leafy greens: Morris MC et al. *FASEB Journal* 2015; 29:260–263S.

Page 272: drinking water before a test increases reaction times: Edmonds CJ et al. *Frontiers Human Neuroscience* 2013; 7:363.

Page 274: swimming is just as effective: Bergamin M et al. *Clinical Interventions in Aging* 2013; 8:1109–1117.

Page 275: Shoot for a minimum of three weekly sessions: Strath SJ et al. *Circulation* 2013; 128:2259–2279.

Page 278: fresh-harvested wild greens were found to have *ten times* as many antioxidants: Trichopoulou A. *Public Health Nutrition* 2004; 7:943–947.

Page 278: spinach has a very high antioxidant capacity: Pellegrini N et al. *Journal of Nutrition* 2003; 133:2812–2819.

Page 279: blackberries contain even more antioxidants: Ibid.

Page 283: the compounds present in this type of chocolate: Buitrago-Lopez A et al. *British Medical Journal* 2011; 343:d4488.

Page 285–86: It can be effective within just a few months' time: Harvie MN et al. *International Journal of Obesity* 2011; 35:714–727.

Page 286: Outdoor exercise, even if moderate like a walk in the park, has a soothing effect: Bratman GN et al. *Proceedings of the National Academy of Sciences USA* 2015; 112:8567–8572.

Page 287: this ancient remedy has long been used to heal wounds: Chimploy K et al. *International Journal of Cancer* 2009; 125:2096–2094.

Page 288: the antioxidants in noni are off the charts: Pawlus AD et al. Noni. In Coates PM et al., eds. *Encyclopedia of Dietary Supplements* (2nd edition). New York: Informa Healthcare, 2010.

Page 289: can have up to three times the antioxidant capacity of red wine: Gil MI et al. *Journal of Agriculture Food Chemistry* 2000; 48:4581–4589.

INDEX

He just wanted a decent book to read ...

Not too much to ask, is it? It was in 1935 when Allen Lane, Managing Director of Bodley Head Publishers, stood on a platform at Exeter railway station looking for something good to read on his journey back to London. His choice was limited to popular magazines and poor-quality paperbacks – the same choice faced every day by the vast majority of readers, few of whom could afford hardbacks. Lane's disappointment and subsequent anger at the range of books generally available led him to found a company – and change the world.

'We believed in the existence in this country of a vast reading public for intelligent books at a low price, and staked everything on it'
Sir Allen Lane, 1902–1970, founder of Penguin Books

The quality paperback had arrived – and not just in bookshops. Lane was adamant that his Penguins should appear in chain stores and tobacconists, and should cost no more than a packet of cigarettes.

Reading habits (and cigarette prices) have changed since 1935, but Penguin still believes in publishing the best books for everybody to enjoy. We still believe that good design costs no more than bad design, and we still believe that quality books published passionately and responsibly make the world a better place.

So wherever you see the little bird – whether it's on a piece of prize-winning literary fiction or a celebrity autobiography, political tour de force or historical masterpiece, a serial-killer thriller, reference book, world classic or a piece of pure escapism – you can bet that it represents the very best that the genre has to offer.

Whatever you like to read – trust Penguin.

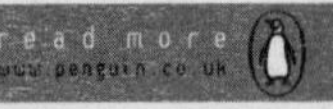